YOU ARE WRONG, MR EINSTEIN

Newton, Einstein, Heisenberg and
Feynman Discussing Quantum Mechanics

YOU ARE WRONG, MR EINSTEIN!

Newton, Einstein, Heisenberg and Feynman Discussing Quantum Mechanics

Harald Fritzsch

University of Munich, Germany

Translated by
Jeanne Rostant

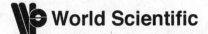

World Scientific

NEW JERSEY · LONDON · SINGAPORE · BEIJING · SHANGHAI · HONG KONG · TAIPEI · CHENNAI

Published by

World Scientific Publishing Co. Pte. Ltd.

5 Toh Tuck Link, Singapore 596224

USA office: 27 Warren Street, Suite 401-402, Hackensack, NJ 07601

UK office: 57 Shelton Street, Covent Garden, London WC2H 9HE

Library of Congress Cataloging-in-Publication Data
Fritzsch, Harald, 1943–
    You are wrong, Mr. Einstein! : Newton, Einstein, Heisenberg, and Feynman discussing
quantum mechanics / by H. Fritzsch.
    p. cm.
    ISBN-13: 978-981-4324-99-1 (hardcover : alk. paper)
    ISBN-10: 981-4324-99-X (hardcover : alk. paper)
    1. Quantum theory--Popular works. 2. Quantum theory--Miscellanea. I. Title.
QC174.12.F755 2011
530.12--dc22

                                                                2010028024

Sie irren, Einstein © 2008 by Piper Verlag GmbH, München

British Library Cataloguing-in-Publication Data
A catalogue record for this book is available from the British Library.

Typeset by Stallion Press
Email: enquiries@stallionpress.com

Printed in Singapore.

CONTENTS

FOREWORD

Harald Fritzsch brings four great scientists back to life for a series of animated conversations about the birth and evolution of quantum theory. They discuss Planck's reluctant introduction of quanta, Bohr's invention of *ad hoc* quantum rules, the joint discovery of quantum mechanics by Schrödinger and Heisenberg, and the fruitful marriage between quantum mechanics and Einstein's special theory of relativity: relativistic quantum field theory. Their conversations reveal how this early work has led us to today's triumphant "Standard Model," which offers an apparently correct, complete and consistent description of virtually all known elementary particle phenomena.

I shall review some aspects of where we are, how we got there and what mysteries remain. All natural phenomena result from the action of four fundamental forces: gravity, electromagnetism and two nuclear forces: one weak, the other strong. Throughout the last decades of his life, Einstein paid scant attention to the nuclear forces. Instead, he strove (with no success) to devise a unified theory of gravity and electromagnetism. One must admit that what we have learned about the nuclear forces has little direct impact on most other scientific disciplines, for which atomic nuclei may be treated as pointlike particles, like electrons, with certain prescribed masses and electric charges. With this idealization, Schrödinger's equation yields a complete description of the atoms of each chemical element. Indeed, it offers the fundamental framework underlying many natural sciences, such as chemistry, biology and geology. Of course, knowing the basic rules does not make these sciences any less challenging, just as knowing the rules of chess does not make one a grand master... but it is a good starting point.

Although gravity is by far the weakest of the four forces, its effects are the most obvious. It explains motion on Earth and in the heavens. It keeps our atmosphere and seas where they belong and our feet to the ground. However, electromagnetism is responsible for almost everything else: how atoms are held together and how they combine together to form such things as molecules, mice and mountains. We ourselves are essentially electromagnetic creatures, as is everything we see, feel, hear, taste, touch or do. Together, gravity and electromagnetism explain almost all features of the world, both great and small. So perhaps Einstein was justified in ignoring the nuclear forces.

But electromagnetism and gravity cannot explain how the sun and the stars shine, nor how the chemical elements of which we are made were created. Without an understanding of the atomic nucleus we never would have faced the promise and perils of nuclear energy. It all began with the discovery of radioactivity in 1897 and the atomic nucleus itself soon thereafter. Scientists discovered nuclei to be made up of two kinds of particles: neutrons and protons that are held together by a strong and short-range nuclear force. Studies of cosmic rays revealed the existence of other seemingly elementary particles: positrons, muons, pions and several varieties of so-called strange particles. The design, development and deployment of ever more powerful particle accelerators has led to the discovery of hundreds of other particles, far too many for all of them to be elementary.

We have learned that most of these particles are not at all elementary. All strongly-interacting particles (called hadrons) are made up of quarks, bound together by the exchange of gluons. For example, the proton is a bound state of two up quarks and one down quark, and a strange particle is simply one containing a strange quark. (However, it is not possible to observe an individual quark or gluon.) The force that holds quarks together, the color force or quantum-chromodynamics (QCD), is based on a gauge theory associated with the mathematical group SU(3). The nuclear force holding nucleons together to form nuclei is but a pale remnant of the color force holding quarks together, just as the chemical force holding atoms together to form compounds is a pale remnant of the electric force holding atoms together.

The last of our tetrad of forces is the weak interaction, which allows protons to become neutrons, thereby enabling the sun to generate energy through the process of nuclear fusion, and long-dead stars to produce the elements of which we are made. It also allows neutrons to become protons in the process of beta decay, one of the three forms of natural radioactivity. Today we know that the weak force is intimately linked (some say unified) with the electromagnetic force. These forces cannot be understood separately, but only together in the form of the electroweak theory, which is based on another gauge theory, one involving the spontaneously broken gauge group $SU(2) \times U(1)$. In this theory, the massless photon is linked to the heavy mediators of weak interactions: W and Z bosons. The electroweak theory has passed every experimental test. Indeed, its construction and development have led to the award of eight Nobel Prizes, including my own.

The electroweak theory, along with QCD, are the two constituents today's triumphant standard model. It is based on the gauge group $SU(3) \times SU(2) \times U(1)$ acting on three families of fundamental fermions, each consisting of a pair of quarks and a pair of leptons. Despite its very many empirical successes, many vexing problems remain unsolved. I conclude this brief essay with a list of some of them:

(i) What about gravity, the one force omitted from the Standard Model? Although this force is too weak to be relevant to elementary-particle phenomenology, it cannot remain a classical force within a quantum world. Thus it is essential that a quantum theory of gravity be formulated. This task may have been accomplished in the context of superstring theory, but as yet this ambitious enterprise is subject neither to empirical verification nor falsification.

(ii) What leads to the breakdown of electroweak symmetry, thereby making the weak force weak and giving most particles their mass? The Higgs mechanism offers a simple explanation, but it leads to a profound theoretical problem. Many ingenious solutions have been proposed (such as supersymmetry and technicolor), but none are convincing. Experiments now being done at the Large Hadron Collider will guide us toward the correct solution, perhaps finding the elusive Higgs boson along the way.

(iii) Cosmologists and astronomers have made many recent and astonishing discoveries: among them, that our universe is flat, that it is expanding ever more rapidly, and that most of the mass in the universe is dark and not made of any known particles. We now know that the universe contains about 70% dark energy, 20% dark matter and only 5% ordinary matter such as is described by the standard model. Thus, two profound problems have been posed: What is dark energy? And, what is dark matter? The former problem seems intractable, but the problem of dark matter can be approached in several promising ways: by producing and detecting it at the LHC; or by observing its interactions as it passes through a detector deep under the ground; or by seeing its indirect effects in the heavens.

(iv) The standard model is more intricate than may appear at first sight. It involves a large number of adjustable parameters whose values must be determined by experiment. For example, the masses and mixings of the various quarks and leptons involve at least 20 independent parameters, most of which have been measured. There seems to be no rhyme or reason for these numbers to be what they are. Surely (we hope) there are physical principles yet to be found that will eventually enable us to compute these numbers from first principles, or at least to find some relationships among them.

Perhaps, in a distant future, a new version of this book will bring back to life those among us or our successors who manage to solve any or all of these problems. Meanwhile, there is much more to be done.

Sheldon L. Glashow
Boston University
Massachusetts, USA

INTRODUCTION

Quantum physics is the science of molecules, atoms and atomic nuclei. Using quantum physics, one has been able to construct lasers, transistors, tunnel microscopes and cell phones; more than one third of the gross national product of our world originates from it. Physicists working on cosmology and astrophysics use quantum physics to study the origin of our universe and the dynamics of stars. It provides the foundations of the physics of elementary particles.

In 1963, I went to Leipzig University to study physics. In the third term I followed lectures on classical mechanics, during which my professor used the excellent textbook of Lew Landau and Evgenij Lifschitz. I had the impression that everything could be calculated exactly, at least in principle. The equations of classical mechanics are quite simple; they determine uniquely the future of a physical system. It is possible to obtain these equations by the principle of smallest action. This principle states that the time evolution of a system is completely determined, if one considers all possible developments and describes them by an action, which is just a simple number. The evolution which is realized in nature is given by the smallest action.

One year later, I was surprised when I followed the lectures about quantum mechanics. I discovered that things were not so clear. Now nothing was certain, there were only probabilities. Those could be calculated exactly, but a degree of uncertainty remained. The principle of the smallest action was not valid in quantum mechanics, and I had problems understanding the details of quantum physics.

A few years later I worked at the California Institute of Technology in Pasadena, where I often discussed with Richard Feynman, who had contributed much to quantum physics. Often Feynman told me: "Nobody understands quantum mechanics, not even I understand it."

Thus, I learned that the phenomena in quantum physics should be understood in a deeper way, and I started to think about the foundations of quantum physics. The fact that one cannot make exact statements in quantum mechanics, but can only state probabilities fascinated me, just like the fact that the stability of atoms and molecules originates from the probability interpretation. If things are exactly determined, as in classical mechanics, there is no stability.

I think that this fascination is not limited to physicists, and with this book I shall try to share this fascination with the reader. In the history of physics, there have been a number of important steps which have provided us with deeper insights into the structure and dynamics of matter.

Isaac Newton realized that the fall of an apple from a tree and the movements of the planets around the sun can be traced back to the same principle, to the gravitational force between massive bodies. Faraday and Maxwell explained why electricity, magnetism and optical phenomena come from the same origin — electromagnetic fields.

Albert Einstein showed in his theory of relativity that space and time belong together. In his theory of General Relativity, he applied his ideas to gravity. Space and time are curved, and the gravitational force is not really a force, but follows from the curvature of space and time. This new view of the gravitational phenomenon creates problems for physicists — as of today, no theory of quantum gravity exists.

The creation of quantum mechanics was also such a step, probably the most important one. Among the achievements of science in the twentieth century, it is the one with the greatest consequences. In physics, there are many phenomena which cannot be understood in the framework of classical physics. Examples are the sizes of atoms, molecules and atomic nuclei, the chemical binding or the stability of atoms and atomic nuclei. Quantum mechanics allows us to understand these phenomena.

Quantum physics was started at the beginning of the twentieth century by Max Planck at the University in Berlin. It was around for two decades, but its foundations were not understood. Then, in only three years, a small group of highly-gifted young physicists, in particular Werner Heisenberg, Wolfgang Pauli and Erwin Schrödinger, created quantum mechanics, the new theory of quantum processes and atoms, based on the ideas of Max Planck, Niels Bohr and Arnold Sommerfeld. In 1928, Werner Heisenberg was 27 years old, Wolfgang Pauli 28, and Erwin Schrödinger 36.

A student of physics learns quantum mechanics with the help of mathematical methods, in particular with differential equations and with functional analysis. Of course, this is not possible in a book such as this one, and so my representation cannot go into great detail. However, I hope that I can introduce the reader to quantum mechanics in such a way that (s)he understands the basic features of the theory. This is possible without mathematics; so mathematical formulae are seldom used here.

In quantum physics, some processes are allowed which, according to the laws of classical mechanics, are not possible. With quantum mechanics, it is possible to calculate these processes. The results are in beautiful agreement with the experimental results.

In quantum physics, a new natural constant is introduced, Planck's action constant, usually denoted by h. It is measured to 6.6×10^{-34} Js (J: Joule = Watt × second). Expressed in macroscopic units like Joule, this constant is very small. This implies that the phenomena of quantum physics do not play any role in the macroscopic world. The name "action constant" comes from the fact that this constant describes an action, which is the product of energy and time. This can easily be understood, since the action of a process is described by an energy, which acts for a certain time. If the time is very small, the action is also very small.

In classical mechanics, the action of a process can assume arbitrary values, but this is not the case in quantum physics. Here, the action can only be an integer multiple of h — the action is always discrete. An action with, for example, $1/3\,h$ is impossible — nature is quantized in units of h. Max Planck discovered this peculiar phenomenon. He proposed that, for example, the energy of an oscillator cannot change arbitrarily, but only in discrete values. Planck applied his hypothesis to the radiation of hot bodies. A piece of iron starts to glow if it is heated. Nobody was able to describe this radiation mathematically, but Planck succeeded and found an equation to describe it successfully.

Albert Einstein used the hypothesis of Planck and proposed in 1905 that light consists of quanta, of particles, which today are called photons. Until 1905, light was considered to be a wave phenomenon. Now it was necessary to treat light both as a particle and a wave phenomenon. Louis de Broglie went even further and proposed in 1923 that all particles are also waves at the same time.

Let me give an example which clarifies the difference between classical and quantum mechanics. In principle, the Earth could have an arbitrary distance to the Sun. But this is not true for the orbit of the electron in the hydrogen atom. It moves on a well-defined orbit; one says that the orbits of the electron are quantized. The electron can jump from one orbit to another if it receives the corresponding energy. In the world of quanta there are no continuous transitions, as in classical mechanics. We shall see later that there is not even a well-defined orbit of the electron, just a probability distribution.

In quantum mechanics, the quantities which describe the movement of an electron in an atom, in particular the location and the velocity of the electron, cannot be measured exactly. There is always an uncertainty which is determined by the uncertainty relation, discovered by Werner Heisenberg. The processes inside an atom cannot be described exactly. One can only give a probability for something to happen.

It is impossible to fix the location and the velocity of the electron exactly. If one wants to know the location with good precision, the velocity is uncertain. But if one wants to know exactly the velocity, the location is very uncertain. The magnitude of the uncertainty is determined by the uncertainty relation, in particular by Planck's constant h.

Uncertainty relations exist also for macroscopic bodies, e.g. for a moving car, but the uncertainties given by quantum physics are so tiny that they can be ignored. This explains why in our macroscopic world, the quantum nature of reality can be completely neglected.

However, this cannot be done in atomic physics. It is precisely the uncertainty which determines the size of the hydrogen atom. Every hydrogen atom has the same size; in the atom, the uncertainty of the location of the electron is given by the atom's diameter, which is about 10^{-8} cm.

Let me now consider a hypothetical hydrogen atom, which is much smaller. The electron is more localized than in a normal hydrogen atom. Due to the uncertainty relation, the velocity of the electron has a much bigger uncertainty, thus on average it moves much faster than in a normal hydrogen atom. The energy of the atom is larger than the energy of a normal hydrogen atom. But in nature there is an important principle: every system tries to be in the state of lowest energy. Therefore, the smaller atom would not be stable. It would emit energy and increase, until it reached the size of the normal hydrogen atom.

We could also consider an artificial atom, which is about a hundred times larger than a normal one. To obtain such an atom, we would have to draw the electron away from the nucleus, thus we have to use energy to produce such an atom. Again, the energy of the new atom is larger than the energy of a normal atom; the larger atom would emit energy and become a normal one. This state is the one with the lowest energy. It is impossible to force the electron to emit more energy. Thus it is the uncertainty relation which fixes the universal size of an atom. This universality is present everywhere in the Universe. A hydrogen atom on Earth is as big as a hydrogen atom on a planet in a distant galaxy.

In the uncertainty relation appears the momentum, not the velocity. The momentum is the product of the velocity and the mass of the particle, e.g. the mass of the electron. Thus the size of an atom depends on the electron mass. If the electron mass were a hundred times smaller than the mass observed in experiments, the size of an atom would be a hundred times larger. If the electron mass were only 0.5 eV, a hydrogen atom would be as large as one tenth of a millimeter.

Due to the uncertainty relation, it is impossible to follow the movement of an electron around the nucleus. In fact, it is impossible to speak about the path of an electron. One can only describe the probability of finding the electron in a certain volume around the nucleus. The hydrogen atom in its ground state is especially simple. The probability distribution does not look at all like an orbit; it is symmetric around the nucleus, and the maximum of the probability is where the proton is located.

The probability distribution is described by the wave function of the electron. This is a function which can be calculated, solving the equation of quantum mechanics. This wave function describes the state of the atom, and often it can be calculated exactly.

If one describes a hydrogen atom using the equations of classical mechanics, the electron would have a certain angular momentum. But in quantum mechanics, the electron in the ground state of a hydrogen atom would have no angular momentum. It does not move around the nucleus on an orbit but vibrates in the vicinity the nucleus.

Another feature of quantum physics is the existence of excited states. If the electron in a hydrogen atom receives energy, e.g. by light radiation, the electron will for a short time be in another state, which has a higher energy. These states are called "excited states", and like the ground state, they have a specific energy. The electron will jump back to the ground state and the energy is emitted in an electromagnetic wave. In quantum mechanics, one can calculate the energy of the excited states.

In 1924, Wolfgang Pauli discovered a new principle, which is now called the Pauli exclusion principle. It says that two electrons in the atomic shell cannot have the same quantum numbers. Using Pauli's principle, it is possible to understand the structure of the atomic shells. In particular, it became possible to derive the periodic system of the chemical elements, developed by Dmitri Mendelejew.

Today there are many applications of quantum mechanics in industry, and modern solid state physics would not be possible without quantum physics. Quantum theory is also very important for the understanding of the atomic nuclei.

Quantum physics is not limited to the microscopic area; it also plays an important role in our daily life. The stability of the desk at which you are sitting at the moment cannot be understood without quantum physics. It is important for chemistry and molecular biology. Only with the help of quantum mechanics is it possible to understand the binding of atoms to molecules.

In quantum physics one has to give up the principle of causality. In classical mechanics, causality is no problem. If one knows exactly the location and the momentum of a particle at a certain moment, the movement of the particle is fixed by the equations of mechanics. In quantum physics this is not true. Due to the uncertainty relation, the laws of quantum physics have a statistical meaning. If one observes a radioactive nucleus, there is no way to predict when the nucleus will decay. One knows only that the lifetime of the nucleus is, e.g., 500 years.

A hundred years ago, the reasons for many phenomena were unknown. Everybody knew that a piece of coal, if heated, became red, then yellow. But nobody could explain this. Why is copper brown and silver white? Why do metals conduct electricity? Why do oxygen and hydrogen atoms bind to produce water molecules? The atoms have a definite size, why?

When physicists started to study atoms, there were many unsolved problems. Let us consider the element neon. Neon is a rare gas without any chemical activity. A neon atom has 10 electrons. But the next element in the periodic table, the element natrium, is chemically very active. Natrium atoms have 11 electrons, one electron more than the neon atom. Thus the chemical properties change drastically if one electron is added. This can easily be understood in quantum physics.

Quantum mechanics works very well as long as the velocities of the particles are relatively small. If they are close to the velocity of light, the theory no longer works, and must be replaced by one which brings together quantum mechanics and relativity theory. This was done by Werner Heisenberg and Wolfgang Pauli, who introduced the relativistic quantum field theory.

The interaction of electrons and photons is in particular described by quantum electrodynamics (QED). Today we describe the interaction of the nuclear particles, the protons and neutrons, by another quantum field theory, which is very similar to quantum electrodynamics: quantum chromodynamics (QCD). This theory describes the interaction of the constituents of the proton, the quarks and the gluons.

Quantum mechanics is for most people a secret science, which only physicists can understand. However, this is not true. The basic principles of quantum physics can easily be understood by everybody. Quantum physics is very important in our modern world, and should be understood by the general public.

Galileo Galilei wrote several books for the general public of his time, for example his famous dialogues, in which he describes discussions between three persons. I have followed his example and written this book in a similar form. The discussions in the book are fictional conversations between Albert Einstein, Richard Feynman, Werner Heisenberg, Isaac Newton and a modern physicist, Adrian Haller from Bern University. Heisenberg was the director of the Max-Planck-Institute in Munich, where I prepared my Ph.D. I worked later with Feynman at Caltech, and after I returned to Europe, he visited me often, either at CERN near Geneva or in Munich.

At the beginning Newton does not understand any quantum physics. But in the course of the discussions he slowly becomes a quantum physicist. Einstein remains skeptical and always has arguments against quantum mechanics. Feynman and Heisenberg defend the modern quantum theory, as does Adrian Haller. The reader, who initially does not understand quantum physics, like Isaac Newton, will learn a lot from the discussions. At the end the reader will be, like Newton, a quantum physicist.

THE START OF QUANTUM THEORY

1

Adrian Haller, physics professor from the university in Bern, went to the annual meeting of the Academy of Sciences of Berlin and Brandenburg, the BBAW, in Berlin. He decided to go by train, from Bern via Basel, Frankfurt, Fulda and Braunschweig to the new main station in Berlin. Haller started to read a book, but became tired; he put the book aside, and after a short time he slept.

In the evening the train arrived at the main station. Haller went to the Friedrichstrasse and took the subway. At the third station he left the subway train and went up the stairs to a big square, called "Gendarmenmarkt". Just across the square was the hotel "Gendarm", where he was to stay. The receptionist greeted him and told him that there were four men who had been in the hotel since the previous day – they had asked for him several times.

"Does one of them have white hair?" asked Haller.

The square called "Gendarmenmarkt" in Berlin.

"Yes, and I have seen him before. His name is Einstein, like the famous Einstein–perhaps he is a relative. Another man speaks only English. His name is Isaac Newton, and he is probably an Englishman or an American. The third is a nice older man, called Werner Heisenberg, who has a South German accent, and the fourth man seems to be an American, a very good looking and very charming man. His name is Richard Feynman."

Haller answered: "Yes, I know these men. In which rooms are they?"

"Mr. Einstein is in room 13. Newton has room 17, Heisenberg is in No. 18, Feynman is in room 20, and you are in room 19, just across the corridor."

Haller went up the stairs to the first floor and entered his room. A few minutes later he knocked at the next door.

"Come in, Mr. Haller, welcome."

Haller opened the door. Three men were sitting on the couch: Einstein, Newton and Heisenberg. Nearby on a chair was sitting Dick Feynman, whom Haller knew very well since his time at Caltech.

Feynman: Adrian, you are no doubt wondering how we knew who was at the door, but this is easily explained – the lady from the reception just called us.

Haller: OK, and I assumed that I should meet you here in hotel Gendarm.

Einstein: A remarkable hotel — it has a wonderful location, in the old part of Berlin. When I lived here, I often walked to the university, through the Tiergarten and passing this market. In two minutes we are at the opera. The Hilton hotel at the Gendarmenmarkt did not exist when I lived here, but the two beautiful churches were here, and the concert hall in between.

Haller: The present building of the Academy of Berlin and Brandenburg, which is called BBAW for short and is basically the previous Prussian Academy of Sciences, is not known to you. When you were a member of the Prussian Academy, it was located in a different building. The present building is also here at the Gendarmenmarkt. I come here often, since I am a member of the Academy. I propose that we go there tomorrow morning. Probably we can get an office there in which we can hold our discussions. You know that we shall discuss quantum physics.

Heisenberg: Before the war I was often in Berlin, also in the Prussian Academy. Tomorrow I shall see the new building.

Haller: Yes, before the war the Prussian Academy was in the building which is now the big library, Unter den Linden Nr. 8. But it is already rather late, and I am tired. I would like to go to my room. I shall see you tomorrow morning, when we have breakfast.

Haller did not go to his room immediately, but went for a walk around the Gendarmenmarkt, passing the two churches. He went to the big street Unter den Linden, passed the opera and the main building of the Humboldt University. There he discovered a plaque showing the name of Max Planck; in this very house Max Planck had worked on quantum physics.

Afterwards he went through the Planck street to the Friedrichstrasse, and then he continued to the Brandenburger Tor. Finally he arrived at the new holocaust monument. It consists of 2,700 pillars of concrete– an impressive monument, referring to the dark time between 1939 and 1945 in Germany.

In the Hotel Adlon near the Brandenburger Tor, he had a drink in the bar. The Adlon Hotel existed before the war, but it was destroyed by bombs. After the unification of Germany, it was rebuilt in the old style. Finally Haller returned to his hotel.

The next morning the five physicists took their breakfast in the hotel. Afterwards they crossed the Gendarmenmarkt to the building of the academy in the Jaegerstrasse. Haller went to the secretary of the president, whom he knew well. She gave him a key to a big office which happened to be empty and which they could use for their discussions.

It was a nice office, with a few chairs, a large table, a blackboard, a couch and a big balcony. They sat down on the balcony; below was the extensive garden, surrounded by the large building of the Academy building.

Haller: A discussion about quantum physics here in Berlin, where quantum physics once started, this is something special. I never thought that this could happen.

Einstein: Yes, we are here at the place where the theory of quantum physics was born. This theory is a product of Berlin, going back to the time before World War One. The first steps were made by Max Planck, only about hundred meters from here, in the university. Yesterday I went there. There is now a plaque at the house where Planck had his office. The people in Berlin are proud of him, it seems. Close to the university there is even a Planck street.

Haller: Mr. Einstein, you also contributed much to quantum theory, especially by the invention of light quanta in Bern in 1905. In 1913 you came here to Berlin, to the center of quantum physics, but I must confess that I never understood exactly why you came here. I guess that Max Planck played a role in this decision.

Einstein: If Planck had written to me, I would have rejected the offer. But he came to Zürich with Walter Nernst. Imagine that: the two great physicists came to me to offer me the professorship in Berlin. When Planck told me about the offer, I said that I would go to the park across the street to think about it. If I should come back after one hour with a bunch of flowers in my hand, it would mean that I had decided to go to Berlin.

I did not think much about the details of the offer itself, since I wanted to go to Berlin, the center of physics in Europe, anyway. Also my dear cousin Elsa was there. But Planck would not get me so easily. I returned without flowers. He was disappointed and told me that I could get two professorships, one at the university and one at the academy, without teaching obligations.

I went again to the park and decided to accept this offer. I bought flowers, and Planck was very pleased. It must be said that he had offered me excellent conditions. The salary was much higher than in Zürich, and higher than the salary of my colleagues in Berlin. But perhaps I should have stayed in Zürich. There I would have had a good, quiet life, even between 1933 and 1945, and I would have not gone to the United States.

Feynman: You would have been professor at the ETH in Zürich – this would have been a good alternative to Princeton. But now let us return to physics.

Heisenberg: I am not sure how we should proceed. I would propose to follow the history. Thus we would discuss first the Planck hypothesis, then the ideas of Niels Bohr about the hydrogen atom, then the hypothesis of Louis de Broglie and the questions about particles and waves. Finally we shall come to modern quantum mechanics. At least for Newton this would be the best way, since he knows little about quantum theory.

Newton: Yes, the historical way seems quite reasonable to me. We start with my mechanics, which I understand very well. But I do have a question about Max Planck. Why did Planck discover quantum physics in Berlin? I know that he was a very conservative man; when he made his discovery, he was already over forty, at an age where one does not make great discoveries. Why was quantum physics not discovered by a very young theoretician, like Albert Einstein, but by the conservative Max Planck, who was quite advanced in age?

Einstein: In 1900 Planck was 42 years old, not really old, but also not a young man. In general, the really new theories in physics are produced by very young scientists, often below thirty years of age. I was only 26 when I introduced the theory of relativity, but Planck was a special man. He was 42, but was to some extent still youthful.

Haller: I know physicists who still produce excellent research work even at an advanced age, for example you, Mr. Feynman. But let us start with our discussion.

Einstein: In 1905 I contributed to quantum physics by introducing light quanta, later called photons. This hypothesis was not liked much by Max Planck. In 1915 very good experiments were carried out

Max Planck

by Robert Millikan in the United States, and he confirmed my theory, for which I was awarded the Nobel Prize in 1922. Later I started to dislike my theory, but the Nobel Prize was very welcome, although I did not profit from it, since the money went to my wife Mileva. For my theory of relativity I would not have received the Nobel Prize – it was too theoretical to the people in Stockholm.

Heisenberg: You would have received the prize for it, but much later. However, it is a fact that you were one of the fathers of quantum theory, a theory which you did not like later. You fathered a child, but you abandoned it – Mr. Einstein, you were not a good father! Others adopted this child: I myself, and also Wolfgang Pauli, Erwin Schrödinger, and in particular Arnold Sommerfeld.

Einstein: You are right, I was a very bad father, but quantum physics is also a bad child. This strange physics is not understood even today.

Feynman: I think one can say for sure, that nobody really understands quantum theory, neither I, nor Mr. Heisenberg, nor Mr. Einstein.

Einstein: I do not want to understand quantum theory. If someone claims he understands it, he is lying.

Newton: Mr. Einstein, please stop criticizing quantum theory. I personally would like to understand it. In my *Principia* I wrote that light consists of small particles. Are your photons my light particles?

Einstein: When I first thought of light quanta, I thought those were your light particles, Mr. Newton. But things are more complicated. Light consists of particles and of waves. This may sound rather crazy to you, but it seems to be true. Let me first explain how I arrived at the idea of light quanta in 1905. This was directly related to the idea of the quantum by Max Planck. Planck introduced the quantum in order to get a theoretical explanation of black-body radiation.

Newton: What is black-body radiation?

Heisenberg: Let me explain this. At the end of the 19th century, physicists discussed which light source was better, the one burning gas, or the one using electricity. The question of what an ideal light source should look like came up. The physicists had an interesting idea. They thought of a cavity, e.g. a ball, containing nothing but radiation. The light inside the cavity can depend only on the temperature of the wall, not on details, e.g. the material of the wall. Both the intensity and the color of the light depend only on the temperature, not on whether the wall is made of metal, wood or stone. Such a cavity is called a "black body". It emits an electromagnetic radiation which is called a "black-body radiation". This radiation has its maximum at a certain frequency. This maximum is shifted upwards if the temperature of the black body is increased. Thus the temperature and the frequency of the maximum are proportional.

Newton: This is strange – I thought that things were more complicated. Thus the radiation in a black body is very simple.

Heisenberg: Yes, it is rather simple. The dependence of the maximum on the temperature can easily be understood. In the equilibrium a black body emits and absorbs the same amounts of radiation. For this reason the radiation does not depend on the properties of the wall. Thus we

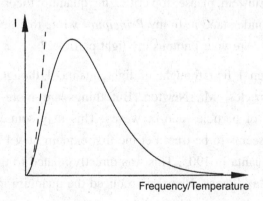

The intensity of the radiation of a black body. The emitted light has a maximum which is shifted upwards if the temperature increases. Classical physics cannot explain this maximum – its prediction is indicated by the broken line.

have an ideal light source, which can be compared to other light sources. One can measure the radiation in a black body if a small hole is made through which the radiation comes out.

Newton: Thus one can measure the dependence of the radiation on the temperature?

Heisenberg: Yes, this can easily be done. But now a problem appears. If the radiation depends only on the temperature, one should expect that the dependence on the temperature is given by a simple mathematical formula.

Einstein: A black body emits no light or electromagnetic radiation with a well-defined wavelength, but there is a distribution of wavelengths. This spectrum of wavelengths can depend only on one parameter, the temperature. Thus one should be able to find a mathematical formula for this dependence.

The radiation laboratory of the Physikalisch-Technische Reichsanstalt in Berlin, in which the experiments were carried out, which supported the quantum hypothesis of Planck.

Haller: Planck tried to find the formula for the dependence of the radiation on the temperature. He found it after making a profound hypothesis. On December 14, 1900 a session of the Physical Society of Berlin was held close to here, at Kupfergraben Nr. 7. Planck gave a lecture, and this lecture is regarded today as the hour when quantum physics was born. Exactly one hundred years later, we had a meeting in the concert hall here at the Gendarmenmarkt.

Newton: Planck must have done something remarkable.

Einstein: Planck's hypothesis changed the development of physics in the 20th century. But for Planck it was not easy. In order to understand black-body radiation, Planck considered the wave theory of light for several years, without success. Then he made a bold assumption. He assumed that the light is not reflected as a wave by the walls of the cavity, but in terms of small elementary units which have a well-defined energy. This energy is fixed by the frequency of light and by a constant, the quantum of action h, introduced by Planck. It is a quantum of the action, and action is the product of energy and time. The product of h and the frequency is energy, since the frequency has the dimension of an inverse time.

This constant h is of crucial importance for quantum physics. Its exact value is $h = 6.62608 \times 10^{-34}$ Js. One Js is the energy of one Joule, acting one second. Note that 1 Js = 1 kg m²/s. Thus in quantum physics there exists a smallest action, different from zero. With this hypothesis Planck was able to find the law for the black-body radiation, given by a simple formula. With this formula he could describe the observations very accurately. The energy of a quantum is given by the simple relation $E = h \cdot v$, where v is the frequency. I have used this relation for my theory of light quanta.

Newton: This constant h is then one of the natural constants?

Haller: Yes, and it is one of the important ones, perhaps even more important than your constant of gravity. Today we also know that the constant h has the same value in distant stars. Even in distant galaxies h has the same value as here on earth. Also billions of years ago h had the same value. In particle physics we do not worry anymore about the constant h – we see it equal to one.

Now we come to a chapter of the development of quantum theory which has to do with you, Mr. Einstein, or more exactly with your theory of photons. Let us first consider the facts. Philipp Lenard, a physicist in Heidelberg, investigated the photoelectric effect. If light is shining on certain metals, electrons are emitted. One would expect the energy of the electrons to be large if one used very intense light. But this was not the case. The velocity of the electrons, i.e. the energy of the electrons, did not depend on the intensity, only on the frequency of the light. Only the number of emitted electrons depended on the intensity, but not the energy of the electrons.

Einstein: I assumed that the light consists of small particles with the energy, given by the formula $h\,v$, where v is the frequency of the light. I was then able to explain Lenard's observations rather easily. If the light is more intense, it means simply that more photons arrive at the metal, but the energy of the photons was the same. Thus the energy of the electrons could depend only on the frequency of the light.

I thought that my theory was a simple application of the idea of Max Planck. But Planck did not think so, and did not like my theory. This is strange, and I never understood it. But I never actually discussed the matter with him.

Newton: But back to the photoelectric effect. If I consider the collision of an electron and a photon, the energy of the electron should depend on the angle between the incoming light and the outgoing light.

Einstein: Yes, if the angle is large, the energy of the electrons is large. Robert Millikan investigated this, and he confirmed my theoretical predictions. But I was not too happy with it. According to my hypothesis, a light ray was a current of many photons. But on the other hand, light was a wave phenomenon. I tried to understand this, but I had no answer. If a photon collides with an electron in a metal, an electron will be emitted. This is a sudden process, and it is impossible to determine exactly when it takes place. In quantum physics the world is not continuous. I did not like this, and Planck did not like it either. But we shall come back to this problem later.

I also have problems with causality. According to our view of the world, nothing happens without a reason. But Planck's quantum hypothesis defines a limit for our causal description of the world.

Haller: We should stop our discussion at this point. I propose that we should now go and have lunch in the nearby steakhouse Churrasco at the Gendarmenmarkt.

Ten minutes later, the five physicists were sitting in the steakhouse. Einstein talked about his life in Berlin, and Heisenberg about his time in Leipzig and in Berlin. After lunch there was no discussion; they went to the street Unter den Linden and visited the old building of the academy, which Heisenberg knew. Then they walked to the Brandenburger Tor and to the holocaust memorial south of the Adlon Hotel. Afterwards they went through the Tiergarten to the Zoo railway station, visiting the nearby church, and returned via Wittenberg Place, where they took the subway back to the Gendarmenmarkt.

ATOMS

On the next morning, the five physicists continued their discussion in the office of the academy.

Heisenberg: Physics deals with the exploration of nature. The aim is to investigate the processes in nature by means of experiments and to describe them by mathematical laws. The Greek philosophers have already described these processes in detail. The history of physics started later, but the ideas of the Greek philosophers were important for the development later on, especially for the atomic theory. The idea that there should be smallest units of matter first occurred among the Greek philosophers about 2500 years ago. Thales, who lived in the sixth century, thought of a unique basic element, which gives the basic matter in our world. A hundred years later, things became more concrete. The philosopher Anaxagoras assumed that there are many different types of matter, whose mixtures produced the many different bodies in our world.

He came rather close to the reality. Today we have about 100 different chemical elements, and they make up everything in our universe.

Then the philosopher Empedokles appeared and proposed that there are only four different elements in the world, earth, water, air and fire. But how can fire be an element? I found this rather strange. The philosophers Leukippos and Demokritos were more concrete. Demokritos introduced the atoms, the smallest elementary objects in the world. The name atom comes from the Greek expression "atomos", which means elementary.

Leukippos introduced the idea of an empty space, in which the atoms are embedded. The empty space was the carrier of the geometry. Demokritos assumed that atoms have no color and no smell. He said: "In reality there are only the atoms and the empty space." Plato mentions in his dialogue "Timaios", that atoms have a regular shape.

The elements earth, water, air and fire are identified with cubes, octahedrons, ikosahedrons and tetrahedrons. For the atoms geometric notions were important. We shall see that the physics of atoms has something to do with geometry. The wave functions of the atoms are simple geometric figures.

It is interesting that in other cultures the idea of an atom did not appear. Only the ancient Greeks had this idea. Unfortunately they never thought of doing any experiments. They wanted to find out things about nature only by thinking, but or course had only limited success. The aversion of the ancient Greeks to any experiments was a big disadvantage. They were philosophers, but not scientists.

Haller: In the year 1417, a manuscript by the Roman poet and philosopher Lucretius with the title *De rerum natura* was discovered in Italy. In this book Lucretius describes the ideas of Leukippos and Demokritos. This book, which was among the first to be printed after the invention of the printing press, was widely distributed in Europe. Lucretius had anticipated very clearly many elements of modern physics.

Heisenberg: The Roman philosophers took over partly the ideas of the ancient Greeks. But after the breakdown of the Roman empire, the Western world declined. Religious fanaticism and superstition dominated for more than a thousand years. The clarity of the Greek thinking became important again in many parts of Europe only at the time of the Italian Renaissance. The modern era of natural sciences started, guided by heroic figures like Kopernikus, Leonardo da Vinci, Johannes Kepler, Galileo Galilei, and also by you, Sir Isaac.

In the 17th century, the atomic science of the ancient philosophers was connected for the first time with concrete scientific ideas. Scientists discovered that the chemical elements like hydrogen, oxygen or copper consist of atoms of the same kind. You, Mr. Newton, went so far to claim that, for example, the firmness of a metal has something to do with the forces, which act between the atoms. This was a good idea, since later it was discovered that it is true.

New ideas came up in the 18th century, proposed mainly by Antoine Lavoisier in France and John Dalton in England. At that time chemical reactions were studied in detail, and it was discovered that water consists of molecules, which consist of an atom of oxygen and two atoms of hydrogen. Later it was found that there are also atoms of electricity, the electrons, and it was discovered that the electrons are parts of the atoms, besides the atomic nuclei. However, not much was known about the structure of the atoms.

Newton: When could something be said for the first time about the size of the atoms?

Heisenberg: This was possible only after the introduction of an interesting number in 1865 by Johann Loschmidt, a physicist in Austria. This number is now called the Loschmidt number. It means the number of atoms for each Mol of matter. One Mol of carbon, for example, is 12 g carbon. The Loschmidt number is given by $L = 6.024 \times 10^{23}$. For the first time one could say something about the radius of an atom, about 10^{-10} m, and about the mass. A hydrogen atom has the mass of about 1.7×10^{-24} g. Thus atoms were indeed very small and very light. It was impossible to weigh a single atom.

Feynman: In the 19th century, spectral lines were discovered. Joseph von Fraunhofer in Munich found the spectral lines in sunlight, but he could not explain their existence. Only later, with the help of quantum mechanics, was it possible to understand the spectral lines. It was found that some substances generate specific spectral lines, i.e. light of a well-defined frequency. Later the chemists used these spectral lines in spectral analysis; chemistry and physics came close to each other.

A discovery by the Dutch physicist Pieter Zeeman was important for the development of atomic physics. He investigated the spectral lines of the atoms, which are in a strong magnetic field. Zeeman found that in the magnetic field the spectral lines were thicker. Later he showed that each spectral line was split into several components. This effect is now called the Zeeman effect.

Professor Hendrik Antoon Lorentz, Zeeman's teacher, interpreted this splitting as evidence that in the neutral atom there are charged particles, which emit light. In a magnetic field a force acts on these particles, which leads to the splitting of the spectral lines. Using the magnitude of the splitting, Lorentz and Zeeman calculated the ratio of the mass of the charged particles and the charge. They also found that the charged particles must have a negative charge, and estimated that the mass of the particles must be very small, only about 1/2000 of the mass of the hydrogen atom.

The charged particles described by Lorentz and Zeeman were discovered in 1897. The English physicist Joseph John Thomson presented his results of experiments with cathode rays at the Royal Society in London on April 30, 1897 and concluded that the cathode rays must be charged particles, which he called electrons. Atomic physics started on this day.

The next important step was made by Ernest Rutherford, a physicist from New Zealand, who worked in Manchester. He and his collaborators were shooting α particles into a thin foil of gold. They observed that sometimes the α particles were strongly deflected. It looked like as if α particles were colliding with particles inside the atoms.

Rutherford was very surprised and tried to explain this phenomenon. In 1911, he found the correct explanation. Almost all of the mass of an atom must be concentrated in the center of the atom. This center must be positively charged — it is the atomic nucleus.

Using this knowledge, Rutherford developed his atomic model. The atom looks like a small planetary system. In the center there is the positively charged nucleus, which carried most of the mass of the atom. The charge of the nucleus is neutralized by the electrons, which circle around the nucleus. The number of the electrons determines the chemical properties of the atom.

The size of the nucleus can be seen in a simple example. Think of the atom as a ball of radius 10 m. Then the nucleus would have a radius of

Ernest Rutherford. He discovered that atoms have an internal structure.

about one Millimeter. Thus atoms are mostly empty, more than 99% of the mass of an atom is concentrated in the nucleus.

The hydrogen atom is the simplest atom, consisting of an electron and a nucleus, which is a proton. The next atom is the helium atom. The nucleus of a helium atom has charge +2, thus there are two electrons in the cloud. Thus one can go on, until one reaches the uranium atom, which has 92 protons in the nucleus, surrounded by 92 electrons.

Heisenberg: Rutherford's model of the atoms has a series of problematic features. An electron, which orbits around the nucleus, should not do that for a long time. The electron oscillates around the nucleus, and oscillating systems generate electromagnetic waves. If such a wave is emitted, energy is lost, and the electron would fall into the nucleus. But we know that this is not the case. Why? Nobody could answer this question.

Furthermore one notices that all atoms have the same structure. Every hydrogen atom has the same radius. Why? In classical physics this cannot be explained. Why is the velocity of the electron such that the distance to the nucleus is always the same, namely 10^{-8} cm?

Bohr solved this problem in a simple way. He assumed that the electrons can orbit the nucleus only on very specific orbits, on which they do not lose energy. Bohr did not know why this is so — he simply assumed it. This was not an explanation, but an interesting idea, and we shall see that Bohr was able to explain many features.

In particular, the experiment of James Franck and Gustav Hertz was interesting for the development of quantum physics. They observed that the electrons inside the atoms and molecules had specified discrete energies.

The solution of the problems in atomic physics was given by quantum mechanics, founded by Max Born, Wolfgang Pauli, Erwin Schrödinger and myself. Quantum mechanics was the theory which allowed us to describe the phenomena of microphysics quantitatively. In the world of atoms, the laws of classical mechanics are not valid. Although the first ideas about the quantum theory were developed by Max Planck in 1900, it took more than 20 years until quantum mechanics was developed, especially by Erwin Schrödinger, Max Born, Wolfgang Pauli, Paul Dirac and myself.

Feynman: But it remained a mystery why quantum theory is so successful in the description of microphysics. I often said: "Nobody understands quantum theory." Niels Bohr used to say that nobody understands quantum theory, unless he is at the same time a bit dizzy in his head.

Einstein: Max Planck had assumed that his oscillators had a specific energy, which can be described by numbers like 1, 2, 3, etc. Bohr assumed that classical mechanics is not valid for the atoms and that the atoms have well-specified energies. In the hydrogen atom the electron can only move on specified orbits, which are called stationary orbits. Classical mechanics would allow infinitely many orbits. In quantum mechanics, this is not allowed. If the electron is on a stationary orbit, no-energy is emitted. This happens only if the electron jumps from one stationary orbit to another. If it has energy $E(1)$ on the first orbit, and energy $E(2)$ on the other orbit, the difference $E(1) - E(2)$ is emitted, i.e. a photon is

Niels Bohr and Arnold Sommerfeld.

sent out. Bohr assumed this without having an explanation for it. But this was exactly what was observed. Bohr's model was not a theory, just a phenomenological model.

The angular momentum of the electron, i.e. the product of radius and momentum, has the unit of an action, the product of energy and time. Bohr assumed that the angular momentum, integrated over an orbit of the electron, is a multiple of Planck's quantum of an action h, i.e. $2 \times h$, or $3 \times h$. Furthermore, Bohr assumed that the atoms exist in well-defined states. If the electrons are in such a state, no energy is emitted. This explains the stability of the atoms.

Bohr was lucky. The hydrogen atom is a very simple atom, and he was able to describe the energy levels of hydrogen with his model very well. It turned out that the electron in the ground state has no angular momentum — this would be impossible in your classical mechanics.

Newton: Let us consider the energy. If we have a circular orbit, then Bohr would set the angular momentum, multiplied by 2π, equal to $h \times n$, where n is an integer, which is called, I think, the main quantum number. Thus one obtains for the energy a constant, multiplied with $1/n^2$, a very simple relation.

Haller: Yes, and the energy differences can be described by a constant, multiplied by $(1/m^2 - 1/n^2)$, where m and n are integers. If m is kept fixed, n is running from $m + 1$ to infinity. The constant is called the Rydberg constant, named after Johannes Rydberg, a Swedish physicist. With the use of lasers one has been able to determine this constant very well: 10973731568525 m^{-1}.

The energy levels give series, which depend on the number m. In 1885 Johann Balmer in Basel, Switzerland, found that the energy levels for the hydrogen atom are given by a constant, multiplied with $(1/4 - 1/n^2)$. This series of levels is called the Balmer series. In this case one has $m = 2$. The series with $m = 1$ was also discovered, but much later, since the light, emitted here, is ultraviolet. The energy levels of this series are described by a constant, multiplied by $(1 - 1/n^2)$. The first state here is the ground state of the hydrogen atom. This series is called the Lyman series, named after the American physicist Theodore Lyman.

Bohr knew the results of the experiments. He found his formula by trial and error. He could not explain why his formula might be right, nor was he interested in that. Only much later, in the second half of 1920s, a better understanding was obtained by Erwin Schrödinger, Max Born and yourself, Mr. Heisenberg.

Let me also mention that the ground state of the hydrogen atom is called the s-state, the first excited state the p-state, the third one the f-state, etc.

Heisenberg: I would like to add that in case of very high excitations the energy differences become very small, and in a good approximation one has a continuum of energy levels. In this region, Newton's mechanics works very well. This feature has obtained a name, it is called Bohr's correspondence principle. It implies that for very high quantum numbers the laws of quantum physics are close to your classical mechanics, Mr. Newton.

Newton: I like this principle. It means that my mechanics works to a certain extent also in quantum physics.

Heisenberg: I should mention that many details of Bohr's model, especially in the case where the electrons do not move on circular orbits, but on elliptical ones, were worked out by Arnold Sommerfeld in Munich, who was my Ph.D. advisor. Sommerfeld has contributed a lot to the atomic model. He should have received together with Bohr the Nobel prize in 1922, but unfortunately that did not happen.

Haller: Today, more than 50 years later, one can read the documents of the Nobel foundation. It was revealed that Bohr himself was against a Nobel prize for Sommerfeld. This is rather strange since the two were good friends.

Heisenberg: I guess Bohr did not want to share a Nobel prize.

Einstein: Let us forget this. Sommerfeld was one of the greatest physicists of the last century, even without a Nobel Prize. He founded the most important school in theoretical physics, with students like Wolfgang Pauli and you.

Heisenberg: Let me return to the hydrogen atom. Let us assume that we make a photo of the atom, not with normal light, but for example with ultraviolet light. Can we see the orbit of the electron?

Newton: Why not? And if we take several photos, we can see the trajectory of the electron inside the atom.

Heisenberg: Mr. Newton, you are wrong. The problem is that the atom is perturbed through the observation, by measuring. Often it will even be destroyed. It is impossible to measure the trajectory of the electron. If a photon collides with the electron, it receives a kick. Thus it is impossible to measure the trajectory exactly, without influencing the electron. We are confronted with the limits of the atomic model of Rutherford. I came to the conclusion that a real trajectory of the electron does not exist.

Newton: Dear Mr. Heisenberg, this must be nonsense, a trajectory for the electron must exist. The electron is a particle, and a particle has a trajectory. Furthermore the electron, which orbits around the nucleus, should also have an angular momentum.

Heisenberg: Sir Isaac, this does not make any sense. In the ground state the hydrogen atom has no angular momentum. I know, in your mechanics this does not make any sense, but your mechanics is not right inside the atoms. The reason is again my uncertainty relation. It is not possible to observe at the same time the location and the velocity of a particle. If we know the location quite well, we know the velocity very poorly, and vice versa. The product of the uncertainty of the location and the uncertainty of the momentum, i.e. the product of mass and velocity, is always of the order of the Planck constant h, never smaller.

Now I would like to introduce the wave function of a particle in quantum mechanics. Let us consider a particle in a potential, for example in a Coulomb potential. It is impossible to determine the exact location of the particle, due to the uncertainty relation. But we can say something about the probability of finding the particle at a specific point. This probability is described by a complex function, which we call the wave function ψ. The product of this function with its Hermitian conjugate function, i.e. $|\psi(x)|^2 = \psi(x) * \psi(x)$, describes the probability. If we integrate this product over a small space volume, we obtain the probability that the particle is in that volume. You see, Mr. Newton, that the quantum mechanics is an exact theory, but it is a theory about probabilities — nothing is certain.

Erwin Schrödinger was the first to calculate such a wave function. The wave function is a solution of a differential equation, which is now called the Schrödinger equation. Observable quantities, for example the momentum of a particle, are obtained by a calculation, involving the wave function. For example, we obtain the momentum of a particle by the derivative of the wave function:

$$p = \frac{h}{2\pi i} \frac{d}{dx} \psi.$$

The different wave functions of an atom are quite different from each other, like the natural vibrations of a string. The various wave functions describe the stationary states of the atom.

Feynman: Schrödinger did not accept at the beginning the probability interpretation. Max Born proposed in 1926 that the square of the wave function describes the probability of finding the particle at a certain location. Schrödinger did not accept this. But today this is common knowledge. Born received the Nobel prize in 1954 for his probability interpretation.

Haller: Back to the hydrogen atom: Bohr described in his model the ground state of the hydrogen atom by a circular orbit. One would expect that the electron has a certain angular momentum. But this is wrong. In the ground state the angular momentum is zero.

Heisenberg: Yes, and we think of the movement of the electron as follows. The electron oscillates around the nucleus. On the average we see a diffuse distribution, the wave function. This function is rotational symmetric, and there is no angular momentum.

But let us return to the beginning of quantum mechanics. I propose that we first look at a very simple system. A particle moves between two walls. This is a rather artificial system, which does not exist in reality, but it is rather simple and can easily be solved.

Newton: You mean a system, in which a particle moves back and forth between two walls. If the particle collides with the wall, it will be reflected. At the reflection point the forces, acting on the particle, become infinitely large. In my mechanics this is a rather simple system. The particle moves back and forth, and the energy is arbitrary.

Heisenberg: In quantum mechanics this is different. Solving the Schrödinger equation is rather simple here. We obtain a wave function for the ground state, i.e. for the state with the lowest energy. The wave function for this state is very simple. It is a sine-function — the wave function is a standing wave. This is easily understood — the wave function is analogous to a string, which is fixed on both walls.

Newton: What do the excited states look like?

Heisenberg: The wave functions of these states are also quite simple. The second state has a wave function, which starts at zero, goes to zero in

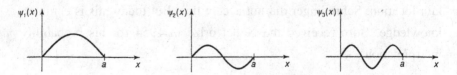

The ground state and the first two excited states of a particle, which moves between two walls.

the middle, and is again zero at the other wall. Also this wave function is a simple sine-function. It has one zero in the middle. The wave function of the next state has two zeros in the middle, etc. If the wave function is zero, it is not possible to find the particle there. In the classical mechanics this would not be possible. The particle moves back and forth, and nowhere should the probability of finding the particle vanish. But in quantum mechanics this is the case. The vanishing of the probability can easily be understood if the electron is described by a wave.

Feynman: We can easily calculate the energies. One finds that the energies are given by Planck's constant, by the mass of the particle and the length of the way the particle moves along. The energy is proportional to the square of n, where n is the number of zeros of the wave function plus one. The first energy is proportional to 1, the second energy to 4, the third to 9, etc. The number n is called the quantum number. The ground state, which does not have a zero, has the quantum number 1, the first excited state has the quantum number 2, etc.

Newton: Yes, this is very simple. The quantum theory is in this case as simple as in my mechanics, perhaps even simpler.

Heisenberg: Yes, sometimes quantum theory is even simpler than classical mechanics. But let us go back to our problem. The quantization of the energy levels is important only if the potential has atomic dimensions. Let us assume that it is big, e.g. 1 cm. In this case the energy difference between two energies, e.g. between the level n and the level $n + 1$, is very small, about $n \cdot 6 \times 10^{-15}$ electronvolt, abbreviated as eV. One electronvolt is the energy, which is received by an electron, if it goes through a voltage difference of one volt. The energy difference between two neighboring states is very small, and in a good approximation we can speak of a continuous spectrum. But if we take a box with atomic dimensions, say with a length of 10^{-7} cm, the energy difference between two neighboring states is $n \times 0.68$ eV. Such an energy difference can easily be measured.

Newton: But something is strange here. The particle could also be at rest in the box. In this case it has the energy zero, and this would be the ground state. But in your calculation the ground state has a certain energy, why?

Heisenberg: This is due to the uncertainty relation. The dimension of the box describes the uncertainty of the location, and once we know this, we can determine the uncertainty of the momentum. The product of the two quantities is of the order of the Planck constant h. Since the uncertainty of the momentum is different from zero, the particle must have some energy, and this is the energy, which Feynman mentioned.

Newton: Again your uncertainty relation fixes the energy, which is different from zero. Quantum mechanics is a strange theory.

Heisenberg: Let me mention something, which Rutherford found in 1919. He identified the nucleus of the hydrogen atom as a new particle, which he called the proton. The nuclei of the other atoms also consist of protons. The number of protons and the number of electron in the atomic cloud is the same.

Rutherford observed that the atomic weight or the mass of the atomic nuclei was always larger than the mass, which he calculated, given the number of protons in the nucleus. In 1920, he postulated that besides the protons, there should be electrically neutral particles, which have about the same mass as the proton. 12 years later, in 1932, the neutral particle, called the neutron, was discovered.

Haller: Gentlemen, it will soon be lunchtime. I propose that we go to the restaurant Lutter–Wegner here at the Gendarmenmarkt. This restaurant has already existed for 20 years and Heisenberg and Einstein will know it. I hope that the quality of the dishes is as high as it was in the 1920s.

The physicists went to the restaurant. After half an hour they found out that the quality of the dishes was indeed very high. They ordered the speciality of the house, veal fillet, and two bottles of wine from the area around the city of Würzburg.

WAVES AND PARTICLES
IN QUANTUM PHYSICS

3

After lunch, the five physicists went to the famous street "Unter den Linden", then continued to the Brandenburger Tor and to the old building of the "Reichstag", where the sessions of the Bundestag now take place. In the Adlon Hotel near the Brandenburger Tor, they made a break at the bar before returning to the office in the academy. Feynman started the discussion.

Feynman: Now we come to a problem which will be particularly interesting for you, Sir Isaac. What is light? At the beginning you supported the wave theory of light, but later, especially in your *Principia*, you proposed the particle theory of light. René Descartes in France worked out many details of your theory. In the 17th century, Francesco Grimaldi in Bologna was in favor of the wave theory of light. Christiaan Huygens discovered subsequently that light exhibits the same phenomena as waves of water — diffraction and interference. Huygens also mentioned the analogy between sound waves and light waves. You started a long discussion with Huygens, but you never reached an agreement.

At the beginning of the 19th century things became more concrete, thanks to the contributions of Thomas Young. He studied medicine and became a universal scientist. Officially he was a medical doctor, but he made many experiments in physics and chemistry. In 1803, he published the interesting result of one of his experiments with light. He invented a device which is known today as the double-slit experiment. Young took a metal plate with two very close narrow slits and observed the light going through the slits. If light were a wave, then light waves would be emitted from each slit. These waves could either enhance each other or cancel each other. On the screen behind the slits there would be dark stripes. If one of the slits were closed, the stripes would disappear. If light consisted of particles, one would not see such stripes.

Young observed the stripes — thus he demonstrated that light was a wave phenomenon. He also determined the wavelength of the red light he was using in his experiment, finding 0.7 micrometer. He presented his results to the Royal Society in London, but many of the scientists did not believe him, and it took another 20 years until it was generally accepted that light was a wave phenomenon. Augustin-Jean Fresnel, an engineer in Paris and member of the Paris Academy, was able to explain many optical phenomena with the wave theory.

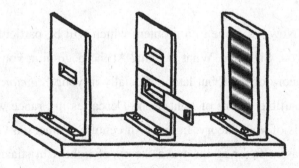

The double-slit experiment. The picture was created by Niels Bohr. The light comes from the left and goes through two slits. If both slits are open, one observes dark and bright stripes on the screen. If only one slit is open, the stripes disappear.

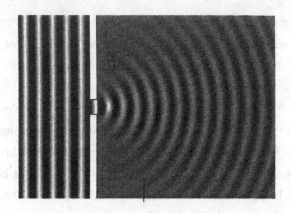

Water waves arrive at a narrow slit, from which circular waves are emitted.

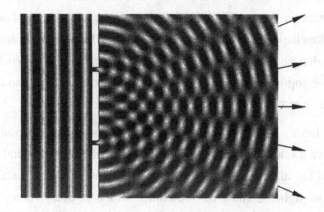

Water waves come from the left and reach the double slit. Circular waves which interfere with each other are emitted from the two slits.

Newton: Thus it seems that I was wrong — light is a wave phenomenon. Waves and particles are quite different. A particle is a point-like object, and waves are big, extended systems. Einstein also introduced the idea of particles of light, like me. Is Einstein's theory also wrong?

Haller: We will return to Einstein's theory later. In the 19th century, scientists in Germany and England developed a detailed theory of light, based on the wave theory. Many optical phenomena were explained, and a new optical industry developed. Especially successful was the Zeiss factory in Jena, founded by Carl Zeiss, Ernst Abbe and Otto Schott in 1846. Abbe was also a physics professor at the University of Jena. Very good optical instruments were built in Jena, for example excellent telescopes or microscopes.

In 1895, Wilhelm Conrad Röntgen in Würzburg discovered X-rays. He did not see any diffraction, and their nature remained unclear until 1912, when Max von Laue in Munich observed the diffraction of X-rays using crystals. Thus the X-rays were like the light electromagnetic waves. In comparison to visible light, the X-rays have very short wavelengths.

Feynman: Mr. Newton, light is only a small part of the spectrum of electromagnetic waves. There is ultraviolet light with a very short wavelength, likewise the gamma rays or the X-rays. Infrared light has a very long wavelength compared to visible light, like radio waves. Visible light has a wavelength slightly larger than the radius of an atom, and is only a very small part of the whole electromagnetic spectrum.

Newton: Let us assume that light is indeed a wave phenomenon. In that case I have a question. A water wave consists of water, a sound wave is a wave in the air, but a light wave consists of what? What oscillates in the case of light? It must be something which is everywhere, here on earth and also in the intergalactic space, a kind of ether.

Einstein: Yes, the 19th-century physicists introduced the idea of a special substance which starts to oscillate when light arrives, and they called it ether. But it must be a strange substance, which is everywhere, a kind of ghost substance, present here on earth and everywhere in the universe, even far away from any galaxies.

Then a series of interesting discoveries was made. It started with the Danish physicist Christian Oersted. In 1820, he carried out an experiment with his students, and he observed something strange. The needle of a compass which had accidentally been placed close to a wire, in which an electric current was flowing, was deflected. Thus he discovered that an electric current generates magnetic effects: electricity and magnetism were related. The next discovery was made by Michael Faraday in England in 1831. He found out that a moving magnet generates an electric voltage in a wire. Today we use this for the generation of electric currents with turbines. Faraday did many experiments with electric currents and magnets. He had the interesting idea that the electric and magnetic phenomena are due to fields which propagate through space and replace the ether.

Electric and magnetic fields are not independent of each other, but are strongly related and influence each other. An electric current, for example a moving electric charge, will generate a magnetic field, and a magnetic field, which changes in time, will generate an electric field. Thus the electric and magnetic fields are closely connected and should be called electromagnetic fields. These fields, which propagate in space, do not need the ether. But even at the beginning of the 20th century physicists were speaking of the ether. In the years following 1860, the English physicist James Clerk Maxwell found the equations which describe the propagation of the electromagnetic fields. They were published by the Royal Society in 1864, and today are known as the Maxwell equations. They are probably the most famous equations in physics.

The Maxwell equations are remarkable. They were formulated at a time when the theory of relativity was not known, but later it turned out that the Maxwell equations were in fact relativistic equations. If he have had lived longer, Maxwell could have discovered relativity theory, but sadly he died at the early age of 48, and I discovered relativity theory in 1905. The Maxwell equations were a complete success even without relativity theory, and even today no changes of these equations are necessary — they seem to be absolutely correct. Your mechanics, Sir Isaac, had to be replaced by relativistic mechanics, but the Maxwell equations remained unchanged.

James Clerk Maxwell (1831–1879), who unified electricity and magnetism. He predicted that there should be electromagnetic waves. They were discovered by Heinrich Hertz in 1888.

I would like to stress again: a magnetic field, which changes in time, generates an electric field, which also changes in time, and this again generates a magnetic field, which changes in time, etc. This is all contained in the Maxwell equations, and you can imagine what this means.

Newton: Yes, it means that the two fields are coupled together. The fields can exist even without any charged particles. The result will be a wave propagating through space, an electromagnetic wave, with a speed, which might be the speed of light.

Feynman: Yes, these are the electromagnetic waves which were predicted by Maxwell. By means of his equations, Maxwell was able to calculate the velocity of the propagation of the electromagnetic waves, and he obtained the velocity of light, about 300,000 km per second. Light is also an electromagnetic wave, but with a rather short wavelength. For example, red light has a wavelength of about 700 nm, 1 nanometer (nm), is 10^{-6} mm $= 10^{-9}$ m.

Electromagnetic waves were discovered in 1888 by Heinrich Hertz, who was a physics professor at the technical university in Karlsruhe. Hertz generated the electromagnetic waves with a transmitter and registered them with a receiver. Hertz's discovery opened up the development of radio techniques, and soon afterwards the first radios were produced. Today we are flooded with electromagnetic waves, especially those coming from television stations. When I look at the programs of our television stations, I often wish that Hertz had never discovered them!

Newton: This is great — light is nothing but an electromagnetic wave. Those waves do not need any ether. My particle theory is wrong — I have to rewrite part of my *Principia*.

Feynman: Sir Isaac, you are too modest. We shall see soon that your theory was not completely incorrect. Yes, light is also an electromagnetic wave, with a rather small wavelength. Radio stations generate electromagnetic waves with rather large wavelengths, about 10 m up to about 100 m in the case of the short-wave radio signals, and even more for the medium-frequency radio waves.

Newton: Now I do not understand Mr. Einstein. You introduced the photons, the particles of light. But if light is a wave phenomenon, this cannot be right. We were both wrong.

Einstein: No, it is not so simple. At the beginning of the 20th century there was a crisis concerning the understanding of light. In the case of the X-rays one could not see any interference. Were the X-rays particles or waves? William Bragg, who received the Nobel Prize in 1915 for his studies of crystals with X-rays, said once: "I teach the particle theory of light on Monday, Wednesday and Friday, and the wave theory of light on Tuesday, Thursday and Saturday." The poor man had to give a lecture every day, it seems. But finally Max von Laue discovered the interference and diffraction of X-rays, using crystals. Those allowed the study of wavelengths, which were about as small as the atoms.

However, there were also experiments which supported the theory that light is a particle phenomenon, especially the photoelectric effect. Here electrons were emitted from a metal if light was directed onto it. The electrons are bound in the atoms of the metal, and a minimum energy is needed to extract them. Heinrich Hertz and Philipp Lenard studied the photoelectric effect in great detail, and found that the kinetic energy of the electrons depends on the color of the light, but not on its intensity. If light were a wave phenomenon, one would expect the opposite. I thought about it, and slowly I found the solution. In 1905, I wrote in my paper that the energy of a light ray is given by a large number of light quanta, which move through space and can only be absorbed fully. Thus the idea of the photon was born. But I was still confused, since the wave nature of light could not be discarded entirely. Nevertheless I received the Nobel Prize in 1921 for my theory.

Later, in 1922, it became clear that my theory was right, shown especially by the experiments of Arthur Compton, who investigated the scattering of light. He found that light, if scattered, changes its frequency. He showed that the energy and momentum of light behave as if it consisted of particles. According to Compton, light is nothing but a stream of photons.

Heisenberg: Yes, but it is not so simple. Light consists of particles, but at the same time light is also a wave. In quantum mechanics this is not a contradiction — it is possible to unite the particle picture and the wave picture of light.

Let me first discuss a simple experiment, which was done by Sir Geoffrey Taylor in England in 1915. Taylor experimented with very faint light. If the particle theory were right, then only one photon at most would go through the two slits per second. Behind the slits Taylor put a film, on which every photon would leave a black point. Taylor looked at the film after a few days. Although only single photons were going through the slits, he observed interference stripes.

Newton: I do not understand this. If only single photons go through the slits, say one photon every second, each of these photons must have gone through the first or through the second slit. Thus I would think that there should be no interference stripes. If many photons go through the slits, then it is different, since the photons could interact with each other and generate the interference stripes in this way. They would be collective effects, due to the many photons involved. Light is a wave, if many photons are present. Single photons are just particles.

Heisenberg: Mr. Newton, this is not right. If one slit is closed in Taylor's experiment, one does not see any interference phenomenon. If a photon goes through one slit, it seems to know, whether the other slit is open or closed. The interference pattern depends on whether one slit is closed or not. It is not a phenomenon which is there if there are many photons involved, but it appears already if only one photon is present. Mr. Einstein, you had many discussions with Niels Bohr about this point.

Einstein: Yes, and I had the opinion that one can figure out for each particle, whether it went through slit No. 1 or slit No. 2, but in this case we have the problem which you just mentioned. Neither I nor Bohr solved the problem.

Heisenberg: Bohr did not have any problem. He always stated that it is not possible to know exactly the path along which the photon went, and to observe the interference stripes. For example, one could have an electron beam perpendicular to the photon beam. If there is an interaction of the photon with an electron close to slit No. 1, one knows that the photon must have gone through this slit. If one knows the path, there is no interference. But it is there if the experiment is done in such a way that the path is not known. Thus we have a choice: either we know the path, in which case no interference is present, or we do not know the path, in which case the interference stripes are there. The interference stripes are an indication of our ignorance concerning the path.

Einstein: Quantum mechanics is a strange theory. I cannot understand it, for me it is a mystery.

Newton: You exaggerate, Mr. Einstein. I find quantum mechanics exciting, something which one does not understand quickly. It is a wonderful theory, and you actually helped to create it at the beginning.

Heisenberg: Quantum mechanics is also a rather precise theory. If we consider a single photon, we can say something about the probability of finding it somewhere. There are no certainties, but we can calculate the probabilities exactly. At the beginning I also had problems with the statistical interpretation. The idea of the probability interpretation was brought up

Max Born, who introduced the probability interpretation of quantum physics. He received the Nobel Prize in 1954.

Erwin Schrödinger (1887–1961). In 1926 he formulated an equation, later called the Schrödinger equation. He received the Nobel Prize in 1933.

by Max Born in Göttingen as follows. The square of the wave function is the probability. A wave function is not a physical field, but some sort of ghost field. The movement of a particle follows the rules of probability theory, but the wave function is exactly determined by the Schrödinger equation. The probability interpretation, introduced by Max Born, was the end for the strict causality in nature. Schrödinger did not accept this, however.

Newton: What about accidents? In our daily life we often speak about an accident. When I stand under an apple tree, it can happen that an apple falls down and collides with my head. I would regard this as an accident. The apple is held by a branch, and when the branch breaks, the apple falls down. Thus the fall of the apple is not an accident, it is a causal process. We call it an accident, since we do not know the details.

Heisenberg: You are right. I speak in such a connection about a subjective accident. We do not know much about it, and for this reason we speak of an accident. But in reality it is not an accident — it is a causal process.

Newton: It could be that the accident which we discuss in quantum physics is of this kind. We do not know any details, and for this reason we speak of an accident. But if one investigates the process further, it looks different. It could be that new parameters exist, which we do not know, some kind of hidden parameters.

Heisenberg: No, I think there are no hidden parameters, and Feynman will think the same way, I guess. In quantum physics, the accident arises since there is no reason for a special event, not simply because we do not know the detailed reasons for an event happening. For example, a particle can fly through this slit, or through the other slit. Since both possibilities are allowed, we obtain an interference pattern. In quantum physics, an accident is not a subjective accident, but an objective one. Nothing is certain — our world is a world of chance, Mr. Newton.

Einstein: I disagree — you describe a casino world, but our real world is not a casino. I do not believe in the probability interpretation of Max Born. God does not play dice, and we are not croupiers.

Feynman: Dear Mr. Einstein — we grow up in our macroscopic world, and this world is not a casino. All processes have a causal origin. Our conceptions were developed in the macroscopic world. But the world of quantum physics is different. Our macroscopic conceptions are no longer useful, once quantum processes become relevant. But since we have no other conceptions, we must use them even in the quantum world, and it happens that this can be done only if the probability interpretation is used. If we do not use it, we have a problem.

Einstein: I do not accept this — I was always unhappy about the probability interpretation of quantum physics. God does not play dice, and the world is not a casino.

Heisenberg: But Niels Bohr often told you that you should stop giving instructions to God. He knows what he is doing, and He will not ask you for advice. God does play dice, since he likes it, and our world is a casino, although I do not like this expression.

Feynman: Mr. Einstein, God indeed plays dice, you should accept this. The probability interpretation becomes important in particular if one considers Heisenberg's uncertainty relation. We can measure the place of a particle rather exactly, but then the velocity is uncertain, and if we know the velocity more precisely, the location is uncertain. The product of the uncertainties of location and momentum is given by Planck's constant. The location and the momentum are two quantities, which are complementary to each other, as Niels Bohr always stressed.

Heisenberg: In our macroscopic world a body has a certain velocity, and it is at a specific location. But in the quantum world this is not true. The quantum world is different, and we cannot grasp it with our macroscopic conceptions.

Einstein: In 1924, Louis de Broglie in Paris proposed something interesting. He knew that light is a wave, but at the same time a particle — thus light is a dual phenomenon. In his Ph.D. thesis, de Broglie generalized this duality to all particles, e.g., to electrons.

The physics professors in Paris were not sure whether this was a good idea, and asked me for a reference. However, I found the ideas of de Broglie very interesting. Experiments soon supported his ideas. Electron beams were scattered on crystals, and already in 1927, diffraction patterns were observed which were similar to the diffraction patterns of X-rays. A few years later one observed diffraction patterns using a beam of sodium atoms.

Now it remained to be understood why not only photons, but also all elementary particles, had this strange dual character. They were particles and waves at the same time. I was not able to understand the details, but de Broglie received the Nobel Prize in 1929. His ideas were important for the further development of quantum mechanics, which is sometimes called wave mechanics.

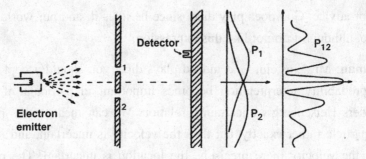

The double-slit experiment. If the second slit is closed, one finds the distribution P(1). If the first slit is closed, one finds the distribution P(2). If both slits are open, one does not obtain the sum P(1) + P(2), but a distribution P(12), which has interference patterns.

Feynman: Indeed, with electrons one can make a double slit experiment just as with light. The interference phenomena are similar to the ones with water waves.

Newton: I do not really understand why one finds these interference phenomena. One can observe every electron, in particular when one knows through which slit an electron is flying.

Feynman: In order to find out through which slit the electrons are flying, you have to scatter light from the electron. But if you do this, there are no interference phenomena. The electrons are normal particles if one observes them, e.g. with light. But the electrons are waves, if they are not observed. This sounds mysterious, but it is easy to understand. The electrons are observed with photons, and if a photon collides with an electron, the path of the electron is changed. In this way, the interference phenomena are destroyed.

Recently it has been possible to carry out double-slit experiments with neutrons, and later with atoms. Even rather large atoms behave like waves in these experiments. Double slit experiments have also been made with fullerenes, which are gigantic molecules, consisting of 60 or more carbon atoms — they look like footballs. Interference pattern have also been observed in the case of fullerenes.

How can it be that an electron is a particle and a wave at the same time? Our classical concepts about the movement of a particle cannot be used in the world of atoms. These concepts were generated by our experience with large objects, which are huge compared to atoms. In order to understand the physics of atoms, new concepts have to be used. Microphysics has different laws from macrophysics. We need both the particle picture and the wave picture to describe an atomic process.

Heisenberg: Let us look at the atomic cloud of the hydrogen atom. With a good camera we can take pictures of the electron. One finds the electron in different places with a certain probability. But the electrons do not have a trajectory. The wave function of the electron is a standing wave. The square of this wave function describes the probability of finding the electron at this location.

As a young student I often wondered about the phenomenon of stability. Why do two hydrogen atoms and an oxygen atom combine to a water molecule? The dynamics of atoms and molecules must be different from the dynamics of classical systems.

In the dialogue "Timaios", Plato wrote that atoms are small triangles, which are able to combine with large geometrical objects, like cubes,

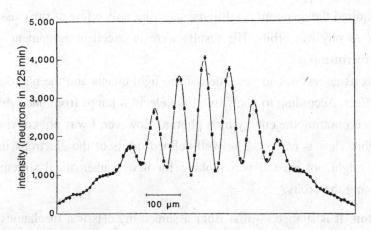

Slow neutrons pass through a double-slit. The solid line shows the prediction of quantum mechanics. The experiment agrees perfectly with the theory.

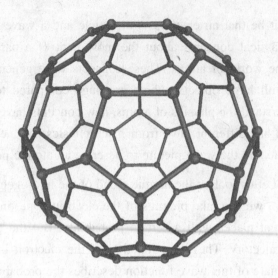

The fullerene C(60) — a carbon atom is located at every corner.

tetrahedrons or octahedrons. Those should be the basic units for the four elements: earth, fire, air and water. I did not believe this, but later I thought that Plato might be right. The atoms and molecules are new mathematical structures, and we had to find out what they are.

Niels Bohr had a simple method for describing the hydrogen atom. He assumed that the electrons move as in classical mechanics, but then he required the quantum conditions, selecting only a few of the classical orbits as physical orbits. His results were in excellent agreement with the experiments.

Let us come back to your idea of the light quanta and the photoelectric effect. According to Niels Bohr, the electron jumps from one orbit to another, emitting the energy by a photon. However, I was not convinced that this view is correct. The well-defined orbits of the electrons in an atom might not exist. I was looking for new mathematical structures, replacing the orbits.

Newton: It is strange — first Bohr assumed my classical mechanics for the description of the orbits, but then he applied the quantum conditions and found that the atoms are stable. The orbits of the electrons here on earth and on a distant planet were identical. Bohr was able to describe the experimental results very well, but nobody understood why this was the case.

Heisenberg: In the spring of 1925, I became ill from hay-fever, and I went to the island of Helgoland. It is in the middle of the sea, far away from blooming flowers. On Helgoland I was able to work, and I replaced the quantum conditions of Bohr and Sommerfeld by new conditions which only had to do with observable quantities, not with the quantum conditions for the fictional orbits of the electrons. I obtained a new theory, and slowly I became accustomed to the new mathematics which I had used in the theory, although I did not understand the details.

When I came back to Göttingen, Max Born told me that I had used matrices in my new theory. I did not know what a matrix was, but I learned it quickly. Matrices have, in particular, the property that the product of two matrices does not commute, like the product of ordinary numbers. The product of two matrices A and B is, in general, different from the product of B and A.

Max Born considered particularly the product of x and the momentum p. He found the relation $xp - px = i\hbar$. This relation is the mathematical basis of my uncertainty relation. Both the location x and the momentum p are now operators, i.e. mathematical prescriptions, not ordinary numbers like in the classical mechanics. Thus the commutator of x and p, i.e. $xp - px$, is different from zero.

Haller: In 2001, on your 100th birthday, a number of physicists went to Helgoland, including myself. We unveiled a monument which is intended to indicate the birthplace of quantum theory. You wrote in your book that you had the idea of the uncertainty relation when you were on a high cliff above the sea, waiting for the sunrise. This cliff does not exist anymore — it was destroyed by a big storm some years ago.

Einstein: Thus the place, where the new quantum theory was born, has disappeared. It would have been even better if the quantum theory had disappeared also!

Newton: Mr Einstein, this is nonsense.

Let us assume that we arrange the following experiment. We produce single photons, for example with a laser diode. Thus the light is very weak, one photon in every minute. The light is directed to a special mirror, which deflects half of the light to the right, the other half goes through. What happens now? A photon is one particle — it cannot be decomposed into several photons.

Heisenberg: I can tell you. The photon will be either deflected to the right, or it goes through the mirror, but it remains one photon. We can put a photon detector at the right-hand side, and another one behind the mirror. The first detector I call the 0-detector, the other one, the 1-detector. Now I start the experiment and note down how often the 0-detector records one photon and how often the 1-detector records one. I obtain a sequence of numbers, say 0 1 1 0 0 1 1 0 1 1 0 1 0 1 0 0 1 0 0 1 etc. This sequence is accidental. It is not possible to calculate what happens after the tenth photon has arrived. But if the sequence has become very long, one has on average as many zeros as ones. The probability to obtain a 0 is 0.50, and the probability to obtain a 1 is also 0.50. Your classical mechanics, Mr. Newton, is an exact theory; everything happens with 100% precision. But in quantum theory things are different — one can only state probabilities. The probability that the 0-detector registers a photon is 50%, and that the 1-detector does it, is also 50%.

Haller: I would like to mention a problem which Erwin Schrödinger discussed a long time ago. Schrödinger did not believe in the statistical interpretation of quantum mechanics. He discussed the following hypothetical experiment. A radioactive atom and a cat are put into a box.

Around the atom is a detector. The atom has a decay probability of 50% in one hour. If the detector registers the decay of the atom, a hammer destroys a phial with a poison, which then kills the cat. The question arises whether the cat is now a quantum cat. If we cannot observe the cat, we should assume that the cat is neither alive nor dead, but a super-position of both possibilities. Schrödinger found this strange: the cat would be something like ("living cat" + "dead cat") / $\sqrt{2}$.

Einstein: This is nonsense. The box could have a window, and I could see whether the cat is alive or dead. Anything in between is not possible. If I were in the box, you would discuss a state like ("living Einstein" + "dead Einstein") / $\sqrt{2}$. What nonsense.

Feynman: I agree with Einstein, I do not see a problem also.

Heisenberg: I also agree — let us not discuss Schrödinger's quantum cat anymore.

Let me mention another interesting property of quantum physics. In classical mechanics, one can distinguish the particles in a system. For example, we can give them numbers or names and can follow their orbits — the particle Hans is there, and the particle Peter is in the other corner.

In quantum theory this is not possible, due to the uncertainty rela-tion. The particles lose their individuality. When I observe an electron somewhere, and a short time later I see another electron, there is no possibility to find out whether the first electron is identical to the sec-ond electron. The particles cannot be identified — it is impossible to give them names. I cannot call one electron Hans, and the other electron Peter. The particle Hans and the particle Peter are the same, they should be called Hans–Peter. I can only observe two electrons. The impossibil-ity to distinguish the particles is a subtle property of quantum physics.

Haller: I propose that we take a break now and go for a walk through downtown Berlin.

The break was quite long. They went to the Alexanderplatz, then to the Pergamon Museum and along the street "Unter den Linden". They stopped for coffee in a restaurant on the Friedrichstrasse, and then went south, to the Potsdamer Platz. Two hours later they returned to the academy.

THE QUANTUM OSCILLATOR

4

Feynman: I propose, Mr. Heisenberg, that you now explain to us another simple system, the harmonic oscillator. I believe you considered that system in particular in the years just before the introduction of quantum mechanics.

Newton: Yes, I know the harmonic oscillator quite well in classical mechanics. I would like to know how it is treated in quantum mechanics. Probably it will not be as simple as in classical physics.

In classical mechanics, the solution of the problem of the harmonic oscillator is very simple. We consider a particle which is fixed on a spring. When this point is moved away from its point of equilibrium, a force will arise, which tries to bring the point back to its point of equilibrium. The force means the product of mass and acceleration,

$$\text{i.e. } m\, d^2 x/dt^2 = -kx$$

(*m*: mass of the particle, *x*: distance to the point of equilibrium, *k*: constant, which describes the strength of the spring, d^2x/dt^2: acceleration).

The solution of this simple differential equation is well known: $x = A \sin(\omega t)$, where *A* is an arbitrary constant and (ω) describes the frequency of the oscillation. It is given by the mass and the constant *k*: $\omega = \sqrt{k/m}$. The sine function has the remarkable property that the second derivative is proportional to the sine function. If we calculate the acceleration, we find that it is indeed proportional to *x*. How is it in quantum mechanics?

Heisenberg: I was a Ph.D. student of Arnold Sommerfeld at the university in Munich, together with Wolfgang Pauli, and I worked mainly on atomic physics. One day Pauli came to me and remarked that it would be very interesting to study the harmonic oscillator in quantum physics. I got interested and started to work on this problem. I succeeded in finding a solution with a kind of new mathematics when I was on the island Helgoland. The solution was very simple. Afterwards, Erwin Schrödinger solved the same problem in a different way, and he found the same result. We shall look at this in more detail. Schrödinger's way is even simpler than mine.

Schrödinger wrote down his equation for the harmonic oscillator. Then he solved his equation, which is a simple differential equation. Let us look at the solution for the simplest case, which is the oscillator in one dimension. I shall first introduce a new parameter, the oscillator length, which is given by the simple formula:

$$b = \sqrt{h/2\pi m\omega}.$$

If I measure the length in units of *b*, the wave function of the ground state, i.e. the state with the lowest energy, is very simple:

$$\phi_0 = \pi^{-1/4} e^{-x^2/2}$$

The energy of this state is:

$$E = h\omega/4\pi.$$

The excited states are described by an index n. The ground state has the index $n = 0$, and the first excited state has the index $n = 1$. It is given by the wave function:

$$\phi_1 = \text{const.}\, e^{-x^2/2} 2x.$$

The const. in front should be chosen such that the integral over the wave function is equal to one. If $n = 2$, the wave function is:

$$\phi_2 = \text{const.}\, e^{-x^2/2}(2x^2 - 1).$$

Let us also consider the case $n = 3$:

$$\phi_3 = \text{const.}\, e^{-x^2/2}(2x^3 - 3).$$

Einstein: For Newton, it will be simpler if he can see the wave functions as curves on the blackboard. I can draw them. The wave function of the ground state is symmetric in x, the first excited state is antisymmetric, the next one is symmetric, etc.

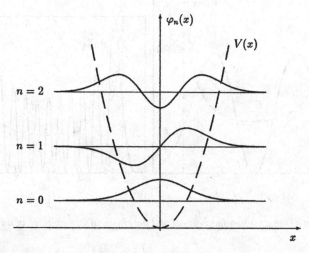

The first wave functions of the oscillator: ground state and the first two excited states.

Feynman: Let us consider the probability for an oscillating particle. In the case of a classical particle, this is easy to calculate. The probability has a minimum at the central point, and it approaches infinity near the two points where the particle turns around. In the figure it is indicated by the broken line. Now we turn to quantum physics and consider the wave function for $n = 1$. The probability is shown by the solid line. At the central point the probability vanishes. Then we have two maxima, and finally the probability goes to zero exponentially.

Let us also consider the case with $n = 10$. Here the wave function vanishes at 10 different points. But when we average the probability, we obtain almost the classical probability. Mr. Newton, you see that classical physics becomes relevant for large values of n. But for small n the deviation from classical physics is significant.

Newton: But when I consider the ground state, I see something strange. In the lowest state the particle should be at rest, and the energy would

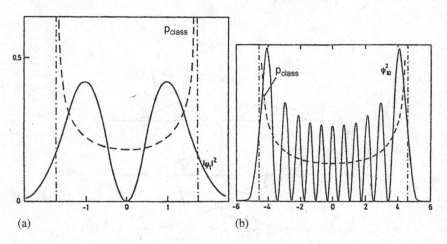

(a)   (b)

A comparison of the quantum-mechanical and the classical probabilities of an oscillating particle.

be zero. But this is not the case. You have also drawn the potential, and one can see that the particle is sometimes in the region which is forbidden by classical physics.

Heisenberg: Yes, in quantum mechanics the particle can be in the region which is forbidden by classical physics. This follows from my uncertainty relation. In classical physics, the lowest value of the energy would be obtained if the particle is at rest. But in quantum physics such a state is not allowed, since it violates the uncertainty relation. If the particle were at rest, its momentum would vanish, and it would be resting at a particular point — this violates the uncertainty relation.

In the ground state, i.e. in the state of the lowest energy, both the momentum and the location of the particle are nonzero, and my uncertainty relation is valid. Also in the ground state the particle moves around. If n vanishes, the energy is given by $h\omega/4\pi$; a smaller value of the energy is not possible. For a classical oscillator all possible energies are allowed, but for an oscillator following the laws of quantum physics, only specific values of the energy are possible.

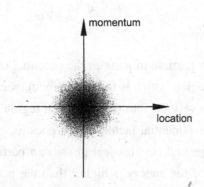

In the ground state of a harmonic oscillator, the momentum and the location of a particle are not fixed due to the uncertainty relation.

Also the other states are quite simple. The energy differences between the states are all equal: $E = h\omega/2\pi$. Thus these differences are twice as much as the energy of the ground state. The harmonic oscillator is a very simple system. The energy increases in equal steps. The eigenvalues of the energy are given by:

$$E_n = \left(n + \frac{1}{2}\right)h\omega/2\pi.$$

As a consequence of the uncertainty relation the particle, being in one of the excited states, can be in a region which is forbidden by classical physics.

Let me now go back to the Schrödinger equation. The energy of the oscillator is given by the equation:

$$E = \frac{p^2}{2m} + V.$$

Here m is the mass of the particle, and V is the potential, which is proportional to the square of x. Thus the Schrödinger equation has the following form:

$$E\psi = -\frac{h^2}{8\pi^2 m}\frac{d^2\psi}{dx^2} + V\psi.$$

We have seen that a particle moving in the potential of an oscillator, can be partially in the region which is forbidden by classical physics. But the particle does not go very far into the forbidden region. The probability is proportional to an exponential factor, which goes to zero very rapidly.

This is true in general. In classical physics, a particle can get over a potential well only if the energy is higher than the potential of the well. But in quantum physics, a particle can move through a potential well if the energy is lower than the energy of the well. This phenomenon is called the tunnel effect.

Recently, we have studied the case where the potential is infinite. In this case the wave function must vanish at the rim of the potential. We can also consider the case where the potential is finite. In this case we have at least one solution of the Schrödinger equation. The wave function is finite beyond the rim of the potential, i.e. the particle can move in the region which is forbidden by classical physics.

Feynman: I would like to mention a technical application of the tunnel effect. Let us take a very sharp needle which is positively charged. We bring this needle close to a metal surface which is negatively charged. If the needle is far from the surface, an electric field exists between the needle and the metal, but there is no electric current flowing. But if the needle is very near to the surface, an electric current starts to flow due to the tunnel effect.

Some of the electrons in the metal use the tunnel effect to jump to the needle. This would be forbidden by classical physics. But due to the tunnel effect, the electric current is flowing. If we move the needle away from the surface, the current vanishes. Thus the tunnel effect can be used to test how far one is away from the surface.

An interesting application of the tunnel effect has been found quite recently. With the sharp needle one can study the metallic surface in detail. Such a device is called a tunnel microscope. It can be used to study in particular very small distances, not much larger than the size of a hydrogen atom. In a measurement a charged needle is moved over a metallic surface. The needle and the surface are not in contact, and there is no electric current flowing. But if the needle is away from the surface by only a few atomic diameters, a tunnel current arises, which depends strongly on the distance.

In a tunnel microscope the distance between the needle and the surface is kept constant, and the needle describes the profile of the surface. A tunnel microscope works however only for metallic surfaces.

Haller: Let me interrupt this discussion. Today I had a phone conversation with a colleague who takes care of the summer house of Einstein in Caputh. We have agreed that we can move tomorrow to this house. Dear Mr. Einstein, I was living in your house some time ago already, when we discussed the theory of General Relativity. I hope you do not mind that we move there.

Einstein: Of course, I have nothing against this. Since we are in Berlin, we might as well go to my house, and you are my guests in Caputh. I would propose that we move there this afternoon.

The physicists went back to the hotel, packed their bags and drove by taxi to Potsdam, and then on to Caputh. After slightly more than one hour they arrived at Einstein's house, Waldstrasse 7. Einstein moved into his room, and Feynman, Haller, Heisenberg and Newton moved into the two guest rooms. Haller and Heisenberg were in one guest room, Feynman and Newton shared the other.

Einsteins's house in Caputh.

Meanwhile it had become late, and they decided not to go to a restaurant, but to have dinner in the house. Einstein went to the supermarket and bought everything needed for a grill party on the terrace. They had grilled lamb chops and drank a lot of beer. The rays of the sun were reflected by the water of the nearby Schwielow lake.

Einstein: Here it is indeed wonderful. Unfortunately I could only live in this house for two summers. In the Autumn of 1932, I went to Pasadena. In January 1933, this criminal Adolf Hitler became Chancellor, and I could not return to Berlin. Eventually I went to Princeton and became a member of the beautiful Institute of Advanced Study. There I often thought of my house in Caputh. I was dreaming of coming back eventually, but it never happened. After the war I could have come back, but I did not want to return to Germany.

Feynman: Why did the Germany elect this Austrian criminal Hitler? Germany was a civilized country, and then this happened, why?

Heisenberg: Nobody knows, it is strange. When Hitler became Chancellor, I thought about leaving Germany, but I did not want to leave my home country. I thought that this affair would be over soon, but I was wrong. It took 12 years, a long time. I should have accepted the professorship at Columbia University which was offered to me. But let us not speak anymore about politics. I am very glad that you are here with us, Mr. Einstein.

The physicists sat on the terrace until midnight. They spoke about politics, especially about the new developments in Europe and about the unification of Germany in 1990.

THE HYDROGEN ATOM

The next morning physicists met on the terrace for breakfast. Heisenberg started the discussion: "Today we shall discuss the quantum aspects of the hydrogen atom. This is the simplest atom, and in classical mechanics it is rather easy to describe. An electron moves around a nucleus, which is just one particle, the proton. Newton could solve this problem easily."

Newton: Yes, this can be easily done. The electron, moving around the nucleus, has a certain angular momentum. The energy is arbitrary, and the problem can easily be solved. The electron moves around on an ellipse, just like the earth around the sun. Now I would like to hear how the problem can be solved in quantum physics.

Heisenberg: First, let me mention that in 1885 the Swiss physicist and high school teacher Johann Balmer discovered the quantized wavelengths of the light emitted by hot hydrogen. He found that the wavelengths followed the simple formula:

$$\lambda = A\left(\frac{n^2}{n^2 - 4}\right).$$

Here A is a constant, and n is a natural number: $n = 3$, $n = 4$, etc. How can these discrete wavelengths be understood?

The Schrödinger equation contains the Coulomb potential, which is a constant multiplied by $1/r$, where r is the distance between the electron and the nucleus. The Coulomb force is proportional to $1/r^2$. Let us consider the lowest states of the hydrogen atom. The hydrogen atom is a symmetric atom — the force, acting on the electron, only depends on the distance to the proton, not on the direction. Thus the angular momentum of the electron is a conserved quantity. In quantum mechanics there are $2L + 1$ different states, which have the angular momentum L. These states just differ by another quantum number m, which varies between $-L$ and $+L$. This quantum number describes the projection of the angular momentum onto the z-axis. It is called the "magnetic quantum number", since it can be measured in a magnetic field.

Newton: The angular momentum is also quantized, I understand — it can only assume discrete values.

Heisenberg: Yes, only discrete values are possible. If we only consider the movement of the electron with respect to the distance to the proton, the problem is reduced to a one-dimensional problem. In this case the force acting on the electron has two components, the electrical attractive force and a force which is fictitious, arising from the conservation of the angular momentum and is relevant only at small distances — the centrifugal force.

If one solves the Schrödinger equation, one finds the energy eigenstates of the electron. For a fixed possible energy there are $2L + 1$ different states. Thus the energy eigenstates are degenerate, and this degeneracy is due to the rotational symmetry of the electrical force. The $2L + 1$ different states have the same angular momentum, but different magnetic quantum numbers. The energy is given by a radial quantum number n. The angular momentum starts with the quantum number $L = 0$, followed by $L = 1$, $L = 2$, etc. If the hydrogen atom is placed into a magnetic field, the degeneracy is lost, and the $2L + 1$ different states are split.

The atomic physicists introduced a description of the states as follows. If the angular momentum is zero, it means that the state is rotational symmetric, and such a state is called an s-wave. If the angular momentum is one, the state is called a p-wave, a state with angular momentum two is called a d-wave, etc. The s-waves are rotational symmetric, thus the wave function depends only on the distance r. But the other states, e.g. the p-waves, have points in certain directions, where the wave function vanishes.

In the ground state, the state with the lowest energy, the wave function is invariant under a three-dimensional rotation, e.g. the probability of finding the electron at a certain point depends only on the distance to the central point. The largest probability is found at a distance of about 5×10^{-9} cm. There is even a certain probability of finding the electron in the central point, where the proton is. In classical physics this would be impossible, but in quantum physics this is allowed.

If one separates the angular momentum, the Schrödinger equation is reduced to a differential equation with respect to the radius r. Solutions of this equation exist only for integer values of the quantum number n: $n = 1$, $n = 2$, etc. For every value of n the angular momentum can have values $0, 1, 2, \ldots (n - 1)$, e.g. for $n = 3$ the values $L = 0$, $L = 1$ and $L = 2$ are possible. These states have the same energy. Thus for a certain possible energy there exist a fixed number of states. If the energy is given by n, then there are n^2 states — for $n = 3$ there are nine states.

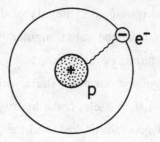

The hydrogen atom, consisting of a proton and an electron.

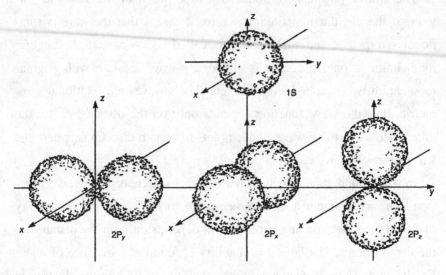

The energy eigenstates of the electron inside a hydrogen atom. The ground state 1S is rotationally symmetric and has angular momentum zero. There are three 2P states with the same energy and the angular momentum L = 1. For those states the probability of finding the electron depends on the direction.

Feynman: Before you continue, let me remind you of the Bohr quantization. Bohr solved the problem simply by postulating that the product of the momentum, e.g. mass × velocity, and the radius should be equal to an integer, multiplied by $h/2\pi$. Bohr was very lucky, since later one found that the Bohr postulate is only exactly valid for the hydrogen atom, due to the high symmetry of the Coulomb force. The Bohr postulate also gives the radius of the state with the quantum number n.

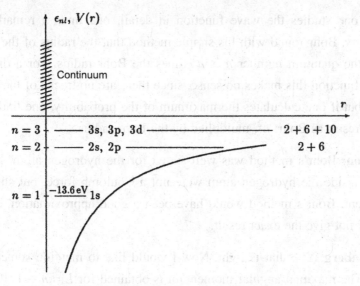

The first eigenvalues of the energy for a hydrogen atom.

It is n^2, multiplied by the radius of the ground state, a very simple result. The energy of this state is $1/n^2$, multiplied by the energy of the ground state. Thus Bohr found the right formula for the observed energy spectrum, e.g. for the Balmer series. With the Schrödinger equation, one can derive the Bohr quantization.

Newton: Please tell us what the exact solutions of the Schrödinger equation are. What are the wave functions of the simplest states?

Heisenberg: You will be astonished — the wave functions are very simple functions of the radius. First, let me give you the *s*-wave as a function of the radius, divided by the so-called Bohr radius, which is 0.529×10^{-10} m. The wave function for the *s*-state is $\Psi = const.\ e^{-r}$ — it is just an exponential. The next excited state with the same angular momentum, e.g. zero, is again an *s*-state, but the wave function now has a zero — it is proportional to $(1 - r/2)e^{-r/2}$. I would like to write down also the wave function of the 2*p*-state — it is proportional to $re^{-r/2}$.

If one studies the wave function in detail, one finds a remarkable property. Bohr found with his simple method that the radius of the state with the quantum number n is n^2 times the Bohr radius. For a diffuse wave function this makes no sense, since there are no tracks of the electron, but if one calculates the maximum of the probability, one finds the same result as Bohr: n^2, multiplied by the Bohr radius.

Newton: Bohr's method was well suited for the hydrogen atom. If the force inside the hydrogen atom were not a Coulomb force, but slightly different, Bohr's method would have been a good approximation, but it would not give the exact result.

Heisenberg: Yes, that is right. Now I would like to mention something else. The maximal angular momentum is obtained for $L = n - 1$. If L is equal to $n - 1$, we have a circular orbit, and when L is less than $n - 1$, we have an ellipse as the orbit. In the ground state with $n = 1$ we have only a circular orbit. For $n = 2$ there are two choices, $L = 0$ and $L = 1$. In the first case one has an ellipse, in the second case a circle.

Newton: Now I understand the hydrogen atom. Bohr's picture is wrong, since it is based on the notion that the electrons move on determined orbits. But the electrons are waves, and these waves could be viewed as vibrations. Every vibration has a certain frequency and a fixed energy. The vibrations are the different states of the hydrogen atom. The wave nature of the electron and the fact that we have states with discrete energies, are directly related. The atom cannot change its state continuously. It must jump from one state to the next.

Heisenberg: Yes, and when Schrödinger calculated the different vibrations of the hydrogen atom, he found that the energies of the various states agreed exactly with the observed energies. This was a great discovery.

Now I would like to mention the series of the spectral lines, for example the Balmer series. Schrödinger calculated the differences of the energies and found for the corresponding wave lengths of the emitted light:

$$\frac{1}{\lambda} = R\left(\frac{1}{n_f^2} - \frac{1}{n_i^2} \right).$$

Here R is the Rydberg constant. The experiments give R = 10 973 731.568 527 (73) m^{-1}. The numbers n_f and n_i are the quantum numbers of the final state and of the initial state. The Balmer series is obtained for $n_f^2 = 4$. Balmer's formula explained the transitions of the electrons to the state with $n = 2$.

The transitions to the ground state were discovered later, for an obvious reason. The photons which are emitted in the transition to the ground state have a very high energy. The spectral lines are in the ultraviolet part of the spectrum. They belong to the Lyman series. The Paschen and Brackett series are in the infrared part of the spectrum. These spectral lines come from electrons, which jump to the states with $n = 3$ and $n = 4$.

Some general remarks about quantum mechanics – this theory gives a reason why the molecules and crystals are ordered. The structure of our world is connected with simple atomic laws, for example the structure of a snowflake or the symmetrical structures of flowers. The reason for this is the same as for the hydrogen atom — all atoms look alike. Our world is full of characteristic structures, but only quantum physics can explain this. Also the stability of atoms can be explained by quantum mechanics. In order to excite the hydrogen atom from the ground state to the next excited state, one needs an energy of at least 10 eV, otherwise the atom remains in the ground state. At a temperature of about 20°C the energy of the atoms is only about 1/40 eV. Thus all the atoms remain in the ground state.

Newton: The hydrogen atom consists of a proton and an electron. How much energy is needed in order to separate the proton and the electron?

Haller: If I take the energy of the proton and electron at rest to be zero, then the energy of the ground state is given by −13.6 eV. This is a rather

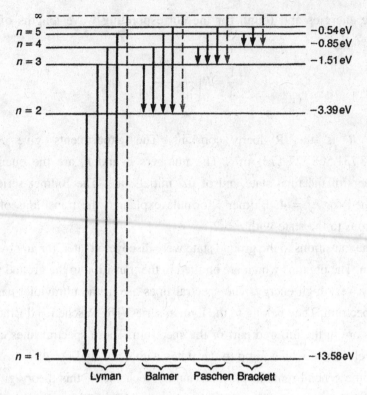

		−0.54 eV
$n = 5$		
$n = 4$		−0.85 eV
$n = 3$		−1.51 eV
$n = 2$		−3.39 eV
$n = 1$		−13.58 eV

Lyman   Balmer   Paschen Brackett

The energy levels of the hydrogen atom. The Lyman series, the Balmer series, the Paschen series and the Brackett series are shown. Only the Balmer series, which describes the transitions of the electron to the state with n = 2, is given by photons of visible light.

small energy. The atoms are not strongly bound, so the electrons can be easily liberated.

Einstein: We know that the proton is not a point-like object, but it has a certain size, about 10^{-13} cm. Thus the equations which describe the atom are slightly changed. Are these changes relevant?

Haller: No, they can be neglected. The effects should be smaller than about 10^{-9} of the energy, which one obtains for a point-like proton.

Newton: The hydrogen atom looks a bit like a planetary system. But there are objects, which come from far away, go through our planetary

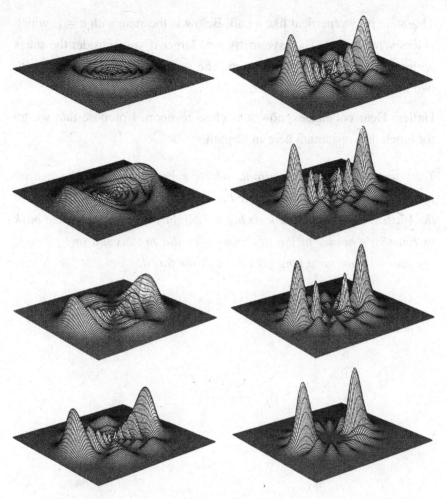

The hydrogen atom with n = 8 and L = 0 (above, left), until L = 7 (below, right).

system and disappear again, like some comets. Do such states also exist in atomic physics?

Feynman: Yes, there are such states, but they are not really atoms. An electron flies close to a proton, and it is deflected. Then it disappears again. The energy of the electron is not quantized — it can be arbitrary.

I would like to show you a picture which describes the state with $n = 8$, a highly excited state. On the left, you see the state with $n = 0$.

This state is symmetrical like a ball. Below is the state with $n = 1$, which is less symmetrical. The asymmetry gets larger if you consider the states with increasing angular momentum. The state with $n = 7$ is a state with two hills opposite to each other.

Haller: Dear colleagues, now it is close to noon. I propose that we go for lunch at a restaurant here in Caputh.

They went to a fish restaurant, which served fish from the nearby Schwielow lake. They accompanied their meal with very good wine from the Unstrut valley, and drank six bottles. Slightly drunk, they walked back to Einstein's house. In the afternoon they did not discuss any physics, but instead went for a long hike through the forest.

THE SPIN — A NEW QUANTUM NUMBER

6

Next morning, immediately after breakfast, Newton started the discussion: "Recently I read that electrons have a new strange quantum number, the spin, some kind of angular momentum. I find this strange, since the electron is supposed to be a point-like object, without any extension, thus it should have no intrinsic angular momentum. But what is the spin, Mr. Heisenberg?

Heisenberg: Yes, the spin is indeed a strange quantum number. It all began with an experiment. In 1922, Otto Stern and Walter Gerlach did an interesting experiment. They were shooting a ray of silver atoms through an inhomogeneous magnetic field, and they observed that the ray was split into two. The atoms of silver have only one electron on the outside, and Stern and Gerlach could observe the electrons. If they had taken a ray of electrons, the same would have happened.

A ray of silver atoms going through a magnetic field and splitting into two rays.

Newton: I guess this means that the electrons must have a property which thus far nobody had thought of. If the electrons were point-like objects without any structure, no splitting would occur.

Heisenberg: Yes, thus the electrons must have another property, and this has to do with a new idea of Wolfgang Pauli.

Pauli assumed that an electron must have a new internal quantum number, which is described by two numbers. With this idea, Pauli was able to describe the electron clouds of the atoms. But he did not know where this new quantum number came from. In 1925, Samuel Goudsmit and George Uhlenbeck introduced an intrinsic angular momentum of the electron to describe the results of the experiments by Stern and Gerlach. This angular momentum was given by $(1/2)(h/2\pi)$, where h is Planck's constant. They called this angular momentum the spin.

Pauli identified his new quantum number with the spin, but it was clear to him that the spin is not a real angular momentum. It was supposed to be an internal property of the electron. Let us consider an electron with the spin 1/2. It has two different states, a state with spin +1/2, and a state with spin −1/2. The square of the spin is like the square of the angular momentum and it is given by the equation. $1/2 \times 3/2 = 3/4$.

Wolfgang Pauli

Newton: An electron has now two states, one state with spin +1/2, and another state with spin −1/2.

Heisenberg: Yes, and the two states can be described by:

$$\left|\frac{1}{2}, +\frac{1}{2}\right\rangle \quad \text{and} \quad \left|\frac{1}{2}, -\frac{1}{2}\right\rangle.$$

The three components of the spin, i.e. the spin in the x-direction, in the y-direction and in the z-direction, can be written as a matrix:

$$S = \frac{1}{2}\sigma.$$

The three spin matrices are given by simple matrices, which are called the Pauli matrices.

$$\sigma_1 = \begin{pmatrix} 0 & 1 \\ 1 & 0 \end{pmatrix}, \quad \sigma_2 = \begin{pmatrix} 0 & -i \\ i & 0 \end{pmatrix}, \quad \sigma_3 = \begin{pmatrix} 1 & 0 \\ 0 & 1 \end{pmatrix}.$$

If one multiplies two of these matrices, then multiplies them in reverse order, and takes the difference, one obtains the same result as with the angular momentum. This is called the commutator.

$$S_i S_j - S_j S_i = i\varepsilon_{ijk} S_k.$$

The spin of the electron is described by two numbers, and the Pauli matrices act on these two numbers. They form a so-called Pauli spinor. The spin of the electron is 1/2 h, i.e. it is half-integer. If the Planck constant is set to zero, it vanishes. Thus the spin is a quantum effect. It has no classical analogue.

An electron can be described by two arrows, one for the momentum, and one for the spin. If we consider an electron at rest, it has, of course, no momentum, and the spin can either point upwards or downwards. If the spin is pointing upwards, the spin wave function is given by [1/2, +1/2], and if the spin is pointing downwards, it is [1/2, −1/2]. The first number is the absolute value of the spin, and the second number is the third component of the spin in the direction of the z-axis.

Haller: I should mention that an arbitrary spin state is usually described by a linear combination, i.e. by the expression a [1/2, +1/2] + b[1/2, −1/2], where a and b are complex numbers satisfying the condition $a \times a + b \times b = 1$. The spin is only with a certain probability p positive, and with probability $1 - p$, it is negative. If $a = b = 1/\sqrt{2}$, the probability that the spin points upwards is 50%, and the probability that the spin points downwards is also 50%.

Newton: Quantum mechanics is a strange theory — nothing is certain, only probabilities exist. Half the time the spin of the electron is oriented upwards and half the time downwards.

Heisenberg: In microphysics our world is uncertain. We describe microphysics with the conceptions of classical physics. It is astonishing that

An electron with the spin upwards.

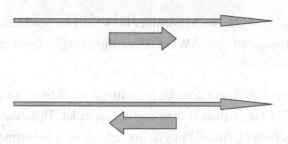

Electrons can be described by an arrow for the momentum and a second arrow for the spin. If the momentum and the spin point in the same direction, the electron is a right-handed electron, and if the spin and the momentum are opposite to each other, it is a left-handed electron.

this is possible, but the compromises we have to accept are the uncertainties, governed by my uncertainty relation.

Feynman: The experiment of Stern and Gerlach showed that the electrons have some kind of intrinsic angular momentum, the spin, and this implies that electrons do have a magnetic moment. This magnetic moment was introduced in 1925 by Samuel Goudsmit and George Uhlenbeck, and they used it to describe the results of the experiments of Stern and Gerlach. The splitting of the beam in a magnetic field was due to the magnetic moment.

The spin of the electrons is also very important for atomic physics. Let us consider again the hydrogen atom. Both the electron and the proton have a certain spin. Both spins might point in the same direction, or they might be opposite.

Newton: But these are rather special cases, in general the spins point in an arbitrary direction, e.g. the spin of the electron in one direction, and the spin of the proton 37 degrees away in another direction.

Heisenberg: Yes, you are right, but when the spin points in some direction, one can always write this as the superposition of two possibilities: spin upward and spin downward. In quantum physics, it suffices to consider only the two possibilities which I mentioned.

Newton: Ok, then one has to consider only the two possibilities: parallel spins or anti-parallel spins. What about the energy? Does it depend on the spins?

Feynman: Yes, the spins act like little magnets. If they are parallel, the energy is a bit larger than if they are anti-parallel. However, the energy difference is minute. Atomic physicists talk about a hyperfine splitting.

Haller: I would like to mention that these hyperfine transitions generate a special light with a wavelength of about 21 cm. This is not a normal light, but an electromagnetic radiation in the microwave region. This radiation is important for radio astronomers. You can see this as follows. The most frequent element in our universe is hydrogen. Not only the stars consist of a large part of hydrogen, but also the big gas clouds in the universe. These clouds cannot be seen directly, but with radio telescopes, since the hydrogen cloud emits radio waves with wavelengths of 21 cm, as the atoms in the cloud frequently make hyperfine transitions. This 21 cm radiation has given us much interesting information about the universe.

Heisenberg: Let us consider another atom, the helium atom. The nucleus of this atom consists of two protons which are held together by the strong interaction between them. However, this interaction is only strong enough if there are also two neutrons in the nucleus. Without the neutrons, the strong electromagnetic repulsion between the protons does not allow the binding of two protons to a nucleus. Thus the nucleus has the structure (*ppnn*).

If we neglect the movement of the nucleus, we have the Schrödinger equation for the two electrons, and a contribution due to the electrostatic repulsion of the two electrons. If we neglect this repulsion, the wave function of the helium atom consists of two wave functions for the two electrons, described by the quantum numbers n, l and m, which can be different for the two electrons.

Let us consider first the ground state. Here is $n = 1$, $l = 0$, $m = 0$ for both electrons. The two spins are anti-parallel. We shall see later that this is related to the Pauli principle. The helium atom which I described is called parahelium, and we can calculate its binding energy. It is about eight times larger than the binding energy of the hydrogen atom. Eight is two times four — the factor four comes from the fact that the nuclear charge is twice as large as the charge of the proton, and the binding energy goes with the square of the nuclear charge. The factor 2 comes from the two electrons. The binding energy of the hydrogen atom is 13.6 eV, thus the binding energy of helium is $8 \times 13.6 = 108.8$ eV. In the experiments one finds a binding energy of only 78 eV. It is smaller due to the electrostatic repulsion of the two electrons, which we have neglected. Besides parahelium, there is also orthohelium. Here the two spins are parallel, thus the spin wave function is symmetric.

Newton: The Schrödinger equation is an equation which depends on the spins. If I consider a particle with spin, the question arises whether the spin is also described by a Schrödinger equation, or perhaps through another equation.

Heisenberg: Mr. Newton, indeed there is another equation. This one was found by another Englishman, Paul Dirac, in the year 1927. It is the famous Dirac equation. We shall look at it soon, but not now; it is lunchtime.

Einstein: Today we shall not go to a restaurant, but down to my garden. I have bought meat, and we will have a barbecue.

FORCES AND PARTICLES IN QUANTUM PHYSICS

7

After lunch, the friends went for a two-hour walk through the forest. It was late when they arrived at Einstein's house; after a short break for coffee, the discussion started.

Heisenberg: In classical physics we have matter, for example a piece of iron, and we have forces, e.g. the electrical or the magnetic force. But in quantum physics a new aspect arises, since it turns out that the forces are connected with particles. I investigated this new feature together with Wolfgang Pauli in the '30s. I expect you are surprised, Mr. Newton — the electrical forces, or more generally the electromagnetic forces, are generated by the exchange of force particles, in this case the photons, the particles of light. These particles fly back and forth between the matter particles and generate the force.

Newton: This sounds interesting to me. It might mean that in the end we have only matter, only particles, and no special forces. The world is simpler this way. We have matter particles and force particles, but both are particles which exist independently.

Heisenberg: Yes, only the matter is relevant. Let me explain how a force is generated in quantum physics. First, let us consider an example in classical physics. We take two boats which are on a lake close to each other. A person in one boat throws a ball to a person in the other boat. This person catches the ball, but receives a push, and throws the ball back to the first person, etc. The net result is that the boats are moving away from each other, as they would if there were a force between them. The electromagnetic force is generated in a similar way — replace the boats by electrons and the balls by photons.

Newton: I understand this example, but it looks to me as if one can generate only a repulsive force by the exchange of balls. How is an attractive force generated?

Heisenberg: Indeed, throwing a ball back and forth will generate a repulsive force. I do not know how to get an attractive force in this case. But we should not take this classical example too seriously. In quantum theory, a force is generated by the exchange of a force particle. This is a quantum phenomenon and has no analogy in classical physics.

The electromagnetic force is generated by the exchange of photons.

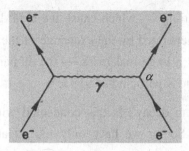

The Feynman diagram for the electrical repulsion of two electrons.

Feynman: I have introduced special diagrams, such as the one you see here, for the description of such quantum processes, and they can also be used to calculate these processes. Here are the two incoming electrons and the exchanged photon. The exchange of the photon leads to a repulsion. If we take an electron and a positron, we have an attraction.

All forces in our universe can be described by the exchange of particles. Photons generate the electromagnetic forces. The strong forces between the quarks, the constituents of the nuclear particles, are also generated by the exchange of particles which are called gluons, and which we shall discuss in more detail later. The weak interactions are also mediated by particles, the weak bosons, the W-bosons and the Z-bosons. The gravitation between massive bodies may come from the exchange of gravitons, but Einstein might not like this interpretation.

I would also like to mention something else. The photons which generate the electromagnetic force do not have a specific mass — their masses vary. Free photons, the photons of visible light, have no mass; they are massless, like the gravitons. The particles which are exchanged have a mass which varies according to the circumstances. Normally this mass is negative. These are not free particles — they are called virtual particles, which live only a very short time. If their mass is large, their lifetime is very short. The concept of virtual particles was introduced in the '30s by you, Mr. Heisenberg, together with Wolfgang Pauli.

The weak interactions — which cause the beta decay of the neutron for example — are generated by the exchange of the W- and Z-particles. These particles have a large real mass — the W-particle has a mass of about 80 GeV and the Z-particle has 91 GeV.

Newton: What is beta decay? Is this connected with forces? What do the W-particles do? So far we have only looked at forces, but not at decays.

Feynman: Beta decay is the decay of the neutron. Neutrons are not stable like the proton, since they decay about 15 minutes after their creation. A neutron decays into a proton, an electron and a neutrino, or more exactly an antineutrino:

neutron → proton + electron + antineutrino.

Neutrinos are neutral particles and relatives of the electrons. Electrons and neutrinos together are called leptons. The decay of a neutron is an effect of the weak interactions and is caused by the exchange of a virtual W-boson.

Newton: Why does the neutron decay, and not the proton?

Feynman: This is easy to understand. The neutron has a mass which is slightly larger than the mass of the proton — the mass difference is rather small, only slightly more than 0.1% of the mass of the proton. Why the neutron mass is larger than the proton mass has remained a puzzle until today. However, in a world in which the proton is heavier than the neutron, we would not be able to live, since the proton would decay into a neutron, and hydrogen would not exist.

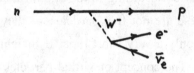

Decay of a neutron into a proton, an electron and an antineutrino.

Newton: Thus our existence depends on this tiny mass difference. But the decay of the neutron is peculiar. One particle changes into three particles. Are the electrons and antineutrinos contained in the neutron?

Feynman: No, they are created in the decay. The mass of the neutron is larger than the proton mass, and therefore it is possible that new particles are created, according to Einstein's relation. For a long time, the details were not known, but now we know that the neutron decays into a proton and a virtual W-boson with a negative charge. The W-boson has a high mass, thus it can be created only as a virtual particle, which immediately turns into an electron and an antineutrino.

Heisenberg: What does one know about the mass of the hypothetical W-particles? You mentioned their masses, but how do you know? The W-particles might not exist.

Haller: No, they do indeed exist. The W-bosons were discovered in 1983 at CERN. At the beginning of the '80s CERN started up a new accelerator, which was able to accelerate protons and antiprotons to high energies and then collide them. The Z-boson was found first in 1983, followed shortly afterwards by the W-bosons. The Z-boson was relatively easy to observe, since it can decay into a muon and its antiparticle, or into an electron and a positron. Its mass is about 91 GeV. The W-bosons often decay into a charged lepton and a neutrino. The neutrinos cannot be observed, and it is not so easy to observe the W-bosons, but at CERN they succeeded. They found the W-boson by observing a charged lepton and a lot of missing energy. The mass is about 80 GeV. The W-bosons were first considered by the theoreticians in the '50s, but it took more than 30 years to find them.

Heisenberg: When I thought about the W-bosons, I had in mind particles with a mass of the order of several GeV, not more than 10 GeV. How are these very high masses generated?

Haller: This is a difficult question, and I cannot give a definite answer. There is a theoretical model, proposed about 40 years ago by

Robert Brout and François Englert in Brussels and by Peter Higgs in Edinburgh. In this model the masses are generated spontaneously by a breaking of the symmetry, which is obtained by interchanging electrons and neutrinos. If this is right, there must exist in our world a scalar particle, usually called the Higgs particle, but its mass cannot be calculated. At the Tevatron accelerator in Fermilab in the United States, the Higgs particle has been searched for without success. CERN's recently-started new accelerator, the LHC, might be able to check this model. The search for the Higgs particle has started, but thus far nothing has been found.

Heisenberg: This scalar particle would be the first elementary scalar particle seen in our world.

Haller: Indeed, so far we have not found any scalar elementary particle. In 1968, I was a young Ph.D. student at the Max-Planck Institute in Munich. At this time we had a regular seminar on the weak interactions, and I had to give a seminar on the W-bosons. In particular I mentioned a model by Sheldon Glashow, in which the W-bosons had a mass of more than 50 GeV. Steven Weinberg considered this theory in 1967 and discussed the introduction of the masses by a spontaneous symmetry breaking, following Higgs and the others. Abdus Salam considered independently the same theory as Weinberg.

The participants in the seminar did not believe such a high value of the mass. But I remember that we had a conversation about the model, and you found it quite interesting. Ten years later, it turned out that the theory was probably close to the truth, and in 1979 Glashow, Weinberg and Salam received the Nobel Prize for it.

Heisenberg: Yes, I remember it well. In your seminar I heard for the first time about this theory with the high masses of the W-bosons, and I found it interesting.

Newton: You mentioned that the force of gravity comes from the exchange of gravitons. Do gravitons exist in our world?

Haller: We do not know — nobody has seen a graviton so far, and I do doubt whether they exist. Mr. Einstein, you introduced the General Theory of Relativity, and in this theory gravity is not a force, but a consequence of the curvature of space and time. In such a case gravitons would not exist, since I cannot imagine that the curvature of space and time is generated by the exchange of a particle. Up to today, nobody has been able to unify the theory of gravitation and the quantum theory. Mr. Einstein, you should look into this.

Einstein: I do not believe that gravitons exist. In my theory, the gravitation follows from the curvature of space and time, as you said, and such an effect cannot be described by particles like gravitons.

Haller: Probably you are right. Today we speak in particle physics about the Standard Model, and in this theory there are matter particles, the leptons, like the electrons, and the quarks, the constituents of the atomic nuclei. These particles have spin 1/2. Between the matter particles, the force particles are exchanged and generate the various forces. The force particles are the photon for the electromagnetic force, the W-bosons and the Z-boson for the weak force and the gluons for the strong force. The force particles have spin 1. The graviton is special — it must have spin 2, if it exists.

Newton: Let me now summarize: particles constitute the matter and also the forces. I like this picture since the world becomes simpler. It would be simpler still if the matter particles and the force particles had the same spin. Matter particles like the electron have spin 1/2, and the force particles, like the photon, have spin 1 — this is strange.

Feynman: A world in which all particles have the same spin would have much less structure than our real world. The complexity of the world has something to do with the spin, and with the fact that there are four different forces in the world: gravitation, the electromagnetic force, the weak force and the strong force.

Heisenberg: Recently I read in a book that attempts are being made to find a new interaction with the new accelerator LHC at CERN, an interaction which is supposed to generate the masses of particles. This would indeed be a new force. Has anything happened so far?

Haller: Yes, the LHC was built mainly in order to find this new interaction, which is related to the Higgs mechanism. The masses of the W- and Z-bosons are generated by a scalar particle, the Higgs particle, which we mentioned earlier. This scalar particle is related to a scalar field, which interacts with itself and with the other particles. The interaction of the scalar field with itself generates a symmetry breaking — the field acquires a nonzero value in the vacuum, which is called the vacuum expectation value. In this way an energy scale is introduced. This energy cannot be calculated, but is related to the Fermi constant, describing the neutron decay. One finds 294 GeV for the vacuum expectation value. The masses of the W- and Z-bosons can then be calculated.

Newton: Why does the photon not get a mass like the W-boson?

Haller: This could be easily arranged. But the theory is constructed in such a way that the photon stays massless. It is unclear whether the mechanism to introduce the masses is really correct, and I have my doubts, but soon the data from the LHC will tell us the truth.

Newton: Now I would like to know more about the strong interactions inside the atomic nuclei. What is known about the strong forces?

Heisenberg: First, I would like to mention that the atomic nuclei consist of nucleons, protons and neutrons; only the hydrogen nucleus consists of just one proton. The protons and neutrons are about 2000 times heavier than the electron.

If two nucleons come close to each other, the strong force binds them together. The strong force binds the nucleons in such a way that a nucleus consists of a certain number of protons and about the same number of neutrons. A nucleus consisting of 20 protons and only three neutrons does not exist. Two protons and two neutrons bind together and form the nucleus of the helium atom, an alpha particle. This is a very stable object.

When the neutron was discovered in 1932, I wondered why the masses of the proton and the neutron are about the same, but their electric charges are different. I proposed that with respect to the strong interactions both particles are identical, and I invented a new symmetry, which I called the isospin. This symmetry describes the interchange of the proton and neutron. The strong interactions respect the isospin symmetry, but not the electromagnetic interactions. The isospin symmetry was the first internal symmetry which was considered in physics, and had nothing to do with space and time.

The electric charge of a nucleus is given by the number of protons inside the nucleus, but the number of neutrons can vary. Thus for a certain chemical element there exist different isotopes. For example, the atom of the element neon always has ten protons, but there are neon atoms with ten, 11 or 12 neutrons in the nucleus.

Some isotopes are not stable and decay after a certain time. If a nucleus has too many or too few neutrons compared to the number of protons, it is unstable. Nuclei which have too many neutrons emit an electron. A neutron changes into a proton and emits an electron. Thus the electric charge of the nucleus increases by one unit. Nuclei which have too few neutrons emit a positron, the antiparticle of the electron, and a proton changes into a neutron. The charge of the nucleus decreases by one unit. These reactions are called beta decays and are caused by the weak interactions.

Besides the beta decay, there is also an alpha decay of nuclei. In this case an alpha particle is emitted. Since an alpha particle consists of two protons and two neutrons, the electric charge of a nucleus decreases in an alpha decay by two units. For example, the uranium nucleus often has 238 nucleons. After the emission of an alpha particle, 234 nucleons remain. They form the nucleus of the element thorium.

Also the element hydrogen has several isotopes. A nucleus consisting of a proton and a neutron is called a deuteron. Atoms with this nucleus are atoms of deuterium, heavy hydrogen. If one has a proton and two neutrons, it is called a triton, the nucleus of the element Tritium. The nucleus of helium normally contains two protons and two neutrons. Sometimes one has only one neutron — this element is called helium-3.

Haller: So far we have discussed the decay of nuclei. Now I would like to change the subject and speak about the opposite, the fusion of nuclei. Two light nuclei can fuse to a heavy nucleus, emitting energy. Again we can use Einstein's formula to calculate the energy — mass is converted into energy.

Einstein: I do not understand. The two nuclei have a positive charge. They repel each other. How could they fuse?

Haller: Yes, it can happen, if the two nuclei collide with rather high velocities. They come close to each other, and the strong interaction fuses the two nuclei. For example, let us take a deuteron and a triton. If they collide, an alpha particle is created, and one neutron is emitted with high energy.

Nuclear fusion takes place in the sun. The energy emitted from the sun is generated by nuclear fusion, and we on Earth profit from it. Without nuclear fusion life as we know it could not exist. For about 50 years physicists have been trying to do something similar on Earth and create a fusion reactor to generate energy. Hydrogen bombs use nuclear fusion, but with bombs one cannot produce useful energy. Fusion reactors are still a hope, but not a reality.

In the south of France, a new test reactor called ITER has been built, and it is hoped to generate some energy for the first time. If the fusion reactors really work, the energy problem on Earth would be solved. The basic material for nuclear fusion, deuterium and tritium, is available on Earth; the water of the oceans contains about 0.015% of heavy water, which contains deuterium. Tritium can easily be produced by the reaction of lithium with neutrons. Lithium is found on Earth in reasonably large quantities. Nuclear fusion has the advantage that it does not produce much radioactive waste.

Einstein: I do not believe that energy can be produced by nuclear fusion, since physicists have been trying unsuccessfully to do so for the last 50 years. But I suppose we shall see.

Haller: Yes, it can happen, if the two nuclei collide with rather high velocities. They come close to each other, and the strong interaction fuses the two nuclei. For example, let us take a deuteron and a triton. If they collide, an alpha particle is created, and one neutron is emitted with high energy.

Nuclear fusion takes place in the sun. The energy emitted from the sun is generated by nuclear fusion, and we on Earth profit from it. Without nuclear fusion, life as we know it could not exist. For about 50 years physicists have been trying to do something similar on Earth and create a fusion reactor to generate energy. Hydrogen bombs use nuclear fusion, but with bombs one cannot produce useful energy. Fusion reactors are still a hope, not a reality. It remains to be seen whether fusion reactors can be built, which produce energy. In that case the energy problem on Earth would be solved. The basic material for nuclear fusion, deuterium and tritium, is available on Earth; the water of the oceans contains about 0.015% of heavy water, which contains deuterium. Tritium can easily be produced by the reaction of lithium with neutrons. Lithium is found on Earth in reasonably large quantities. Nuclear fusion has the advantage that it does not produce much radioactive waste.

Haller: Now let us finish our discussion for today. Tonight we shall not have time to discuss the details of the strong interactions, which are directly connected with the quarks, the constituent of the nucleons. But soon we shall discuss the quarks in more detail.

Now we have to decide where to have dinner tonight. I propose that we go to a restaurant nearby, the "Kavaliershaus" in Caputh.

Shortly afterwards they walked down to the sea and along the shore, until they reached the big restaurant in the center of the village.

THE PERIODIC TABLE

8

The next morning it was raining, and so the physicists met in the living room in front of the fireplace. Einstein brought some wood, and they made a fire.

Heisenberg: Chemists know that the chemical elements can be classified in a periodic table introduced by Dmitri Mendelejew (1834–1907). He did not know much about atomic physics, but he proposed his table which became a big success in chemistry. When he introduced his table, Mendelejew classified the 63 chemical elements, which were known around 1869. He found that the elements with similar chemical properties had quite different atomic masses, and elements whose masses were similar had very different chemical properties. For example, the rare gas neon has ten electrons, and the next element, sodium, with 11 electrons, is a very aggressive element. Mendelejew discovered that the elements could be grouped together in families, and in this way he came to derive his periodic table.

Can we derive the periodic table from quantum theory? Niels Bohr investigated this by studying the filling up of the electronic shells of atoms. If we go from one atom to the next, the nuclear electric charge increases by one unit, so correspondingly one electron has to be added in the shell. But Bohr could not understand why in the lowest energy shell there could only be two electrons.

In 1925, Wolfgang Pauli studied in particular the atoms of the alkali elements. He was able to understand these elements only if a new principle was introduced, which was later called the Pauli Exclusion Principle, and for which he received the Nobel Prize in 1945. It is very important for the understanding of the periodic table.

Niels Bohr remarked in 1912 that it was strange that the electrons in an atom are not all in the lowest state of energy. If this were the case, all elements would have similar chemical properties. The Pauli principle does not allow the electrons to be all in the lowest state. It affects all particles with half-integer spin, e.g. the protons or the electrons. The principle states that two particles with half-integer spin can never be in the same state. In an atom two electrons could be in

Dmitri Mendelejew

the same state, if the spin is ignored, but then the two spins must be opposite to each other.

The Pauli principle is relevant also for the light elements. (Hydrogen has only one electron, thus the principle is irrelevant.) The next element is helium, whose atom has two electrons in the shell. The lowest shell can only have two electrons, due to the spin. Thus in the helium atom the spins of the two electrons are opposite to each other. The shell is now complete, and more electrons are not allowed. This explains why helium is a rare gas and does not bind with other elements to form molecules. The following element is Lithium. Its atom has three electrons, but the third electron cannot be in the lowest shell, due to the exclusion principle — it is in the second shell.

Feynman: Pauli had strange ideas, but they were right. Why the Pauli principle has something to do with our world, only God in heaven knows. It is a peculiar principle. The spin dictates that the particles cannot be in the same state. If the electrons were particles without spin, the physics of the atoms would be quite different. All electrons would be in the lowest shell.

Heisenberg: Yes, atomic physics would be completely different. But let me remind you that the spin is not an additional property of the electron, which could be removed. Particles with spin are described by an equation which has some similarity with the Schrödinger equation, introduced by Paul Dirac in Cambridge in 1928, and called the Dirac equation. It was shown much later that this equation is only consistent with Einstein's theory of relativity if the Pauli principle is obeyed. The proof is difficult, and we shall not discuss it here.

Newton: Ok, let us accept this. The Pauli principle seems to lead to the fact that the structure of atoms with many electrons is rather complex. In the lowest state, which is an s-wave, only two electrons can be accommodated, with opposite spin. The next electron has to be in a p-wave state, etc.

Heisenberg: Yes, in this way the ground states of the atoms are rather complex systems. Without the Pauli principle, our world would be quite simple. All electrons would be in the ground state, in an s-wave, and all elements would have about the same properties. The world would be rather boring, but nobody would be around to register this.

The Pauli principle explains, for example, why the rare gas neon with ten electrons is very different from sodium with 11 electrons. The sodium atom has a single electron outside, and for this reason sodium is chemically very active, unlike neon. The chemical properties of the elements change abruptly. The outside electron of the sodium atom can easily be moved to another atom, e.g. to a chlorine atom. Thus the sodium and chlorine atoms are bound together and give a molecule of salt: $NaCl$.

Feynman: If one knows the number of electrons, one can predict the chemical properties of the element. For example, at normal temperature silver is solid and nitrogen is a gas. Let me also mention the element plutonium. It does not exist on earth, since it decays with a half-time of about 40,000 years, but it was produced in nuclear reactions. It was known that a plutonium atom has 94 electrons. Based on this knowledge, physicists were able to predict that plutonium would be a brown metal. In Los Alamos a cubic millimeter of plutonium was produced in 1945, and indeed, it turned out that plutonium was a brown metal.

Heisenberg: Let me now describe the features of the atoms which are important for the periodic table. The simplest atom is that of hydrogen, where the single electron is in a $1s$-state. The next atom is the helium atom, which has two electrons in the shell. Both electrons are in a $1s$-state. But the binding energy of the electrons is much larger than for the hydrogen atom, since the electric field is stronger, due to the two protons in the nucleus. The helium atom has a complete shell, filled with the two electrons. Thus helium is chemically neutral — it is a rare gas. The two neutrons in the nucleus are needed to have a rather stable nucleus. A rare isotope is helium-3, which has a nucleus with two protons, but only one neutron.

The next element is lithium, with three electrons. The third electron is in a $2s$-state, since the $1s$-state can only have two electrons, due to the Pauli principle. Then the element beryllium follows. Its atom has four electrons, two of them in the $2s$-state. The next elements are boron (B), carbon (C), nitrogen (N), oxygen (O), fluorine (F) and neon (Ne). The neon atom has ten electrons, two in the $1s$-state, two in the $2s$-state, and six in a $2p$-state. The number of electrons with the quantum number 2 is eight. The second shell is completely filled, thus neon is, like helium, a rare gas.

Next we have sodium, whose atom has one electron in a $3s$-state. This electron starts a new shell, implying that sodium is chemically very active. Then things continue, until we again reach a shell which is completely filled. This happens when 18 electrons are involved. We have the element argon (Ar), which is again a rare gas.

The next complete shell is reached if we have 36 electrons, and this happens for the element krypton (Kr). The complete shell after that is obtained if there are 54 electrons: we again have a rare gas, the element xenon (Xe). The last known complete shell is reached for 86 electrons, the rare gas radon, which is radioactive and has a half-life of only about 3.8 days.

The most well-known element beyond radon is uranium, which has 92 electrons and 92 protons; the number of neutrons can vary between 141 and 148. The uranium found on earth has mostly 146 neutrons — it is called U-238. Uranium-238 is not stable and decays into thorium by emitting an alpha particle. However, its half-life is large — about 4.5 billion years.

Newton: I am astonished that for the light elements, for example helium or carbon, the number of neutrons is equal to the number of protons. But this is not true for the heavy elements, like gold or uranium. The uranium atom has 146 neutrons, but only 92 protons. Is this understood?

Heisenberg: This is connected with the nuclear strong force. An atomic nucleus consisting of just protons cannot exist, since the electric repulsion of the protons would not allow a nucleus to be formed. In a real nucleus, the strong force between the protons and neutrons keeps the nucleus together. The strong force between the protons would not be strong enough, but with the neutrons there is no problem. For the light elements the same number of neutrons and protons is sufficient to have a stable nucleus. But for the heavy elements one needs more neutrons, since the electric repulsion is stronger. A uranium nucleus with 92 protons and 92 neutrons cannot exist, but with 146 neutrons it is ok.

Haller: In any case it is possible to understand the periodic table by means of quantum mechanics. For the first time quantum physics contributed to progress in chemistry. As a physicist I like to say that chemistry is a sort of applied physics.

Periodic table of the elements.

Heisenberg: I agree with you — chemistry is a part of physics. My colleagues in chemistry do not like such a statement.

Haller: I think that chemistry is also an interesting science, but certainly physics has contributed a lot towards the understanding of the periodic table.

I would like to mention something which would be interesting in particular to Einstein. In 1925, you considered the problem of what happens to a many-particle system if it is cooled down. Can the particles lose all their energy? This is possible only for a particular kind of particle, the Bose particles.

Newton: What is a Bose particle?

Feynman: The physicist Satyendra Bose in India proposed in 1924 that there are particles with a symmetric statistics, e.g. the photons, and particles with an antisymmetric statistics, e.g. the electrons. Particles of the first kind were later called Bose particles or bosons, and particles of the second kind were called Fermi particles or fermions. The wave function of two bosons must be symmetric under the exchange of the two particles, and the wave function of two fermions must be antisymmetric. Fermions have a non-integral spin, e.g. 1/2, 3/2 or 5/2, and bosons have an integral spin, e.g. 0, 1 or 2. Our stable world consists only of fermions and of photons.

Newton: Another question: most of the atoms do not exist as isolated systems, but are bound to molecules. For example, a hydrogen atom is normally bound with another hydrogen atom to form a hydrogen molecule. How does the binding of the two atoms proceed? The electrons in the shell repel each other, and I do not understand why there is a binding.

Feynman: The quantum mechanics of the atoms is quite simple, but the molecules are not simple systems. Very big molecules, which are studied for example in biochemistry, cannot be understood with the Schrödinger equation alone. The chemists have developed approximate methods to describe these molecules, which sometimes work well, sometimes not.

Newton: I am not interested now in these big molecules, just in the hydrogen molecule. Why do the two atoms form a molecule? A hydrogen atom is electrically neutral, thus the electrical forces are not relevant.

Feynman: Let me simplify the situation and let us consider a hydrogen molecule which is positively charged. It consists of two protons and one electron. The two protons repel each other, and the electron moves around the two protons. One can calculate the energy of this system. The distance r between the two protons is kept fixed, and we now calculate the energy. It is sufficient to calculate the lowest energy level. This energy is a function of the distance r. For a certain distance a minimum of the energy is obtained; this happens at a distance about 1.5 times the Bohr radius. A hydrogen molecule is very similar. Instead of one electron we now have two electrons, which circle around the two protons.

Newton: If the two protons come very close together, we would have a helium atom without neutrons. But a hydrogen molecule and a helium atom are different systems.

Feynman: Yes, but there is no transition between the two systems, since the two protons repel each other. The two neutrons are needed to form a nucleus – the strong interaction between the two neutrons and two protons is able to glue the four nucleons together to a nucleus. With two protons alone this does not happen. Nevertheless, a hydrogen molecule and a helium atom are similar. In the helium atom we have the nucleus with two protons and two neutrons, and in the shell there are two electrons. In the hydrogen molecule there are two protons, which are separated by a certain distance, and in the shell again two electrons. The wave functions of the electrons for the helium atom and for the hydrogen molecule are quite similar.

Let me say something about the spin. The spins of the two electrons in the hydrogen molecule are opposite to each other, as for parahelium. If two hydrogen atoms come together, they only form a molecule if the spins are opposite to each other. If the spins are parallel, the atoms repel and do not form a molecule.

Heisenberg: The atoms form molecules, since they try to achieve a state which is similar to a rare gas atom. An example is salt, the molecule NaCl. The chlorine atom takes one electron from the sodium atom and achieves a shell structure like the rare gas argon. The sodium atom with one electron less has a shell structure like the rare gas neon. The electrostatic force between the two systems leads to the formation of the molecule. In a similar way one can understand other molecules, e.g. the water molecule.

Haller: Now it is lunch time. I propose that we go to eat in Potsdam.

They drove by taxi to Potsdam and went to the restaurant "Zur historischen Muehle", which Haller knew. After lunch, they visited the downtown area of Potsdam and "Sanssouci", the castle of King Frederick II of Prussia.

Quantum Theory and the Relativity of Space and Time

9

Next morning the physicists met again on the terrace. Heisenberg started the discussion.

Heisenberg: Let me come back to the spin of the electron and to a discovery made by Paul Dirac. Dirac tried to combine Einstein's theory of relativity and quantum mechanics. In 1928, he found an interesting equation, which is today called the Dirac equation.

Let me explain how he derived his equation. He was trying to find a relativistic version of the Schrödinger equation, but there is a problem: the Schrödinger equation is non-relativistic. Space and time are used in this equation in a different way. For example, let us consider the Schrödinger equation for a free particle. The time derivative of the wave function is related to the space derivative given by $(p^2/2m)\Psi$. Note that the momentum p is proportional to the space derivative of the wave function.

Paul Dirac

Dirac tried to change the Schrödinger equation in such a way that it is consistent with relativity. In the Schrödinger equation, the second derivatives appear with respect to the space, but only the first derivative with respect to the time. Space and time are different. One could replace the first time derivative by a second time derivative; then one obtains the Klein–Gordon equation. This equation has problems which I do not want to explain now.

Dirac had another idea. He replaced the second derivative with respect to the space by the first derivative.

Newton: I do not understand. The first derivative with respect to the space shows in a particular direction. It is a gradient, in particular a vector. The second derivative is not a vector. It appears as the sum of the three second space derivatives in the Schrödinger equation.

Feynman: Yes, you are right. Dirac avoided the problem with the gradient. He multiplied the four derivatives, three for the space and one for the time, with specific matrices. These matrices are known today as the Dirac matrices. They are 4×4-matrices, and the symmetrized products have specific properties. The Dirac matrices are closely connected to the three Pauli matrices. But details should not be discussed now. The Dirac matrices are 4×4 matrices, and the wave function is not a simple function, but consists of four different functions.

Here are the four Dirac matrices — the index i is related to the space direction. The index 0 describes the time.

$$\gamma^i = \begin{pmatrix} 0 & \sigma^i \\ -\sigma^i & 0 \end{pmatrix}, \qquad \gamma^0 = \begin{pmatrix} 1 & 0 \\ 0 & -1 \end{pmatrix}.$$

Newton: But why are there four wave functions? I thought that the electron has two wave functions, due to the spin.

Heisenberg: Yes, the two components describe the spin. The other two components also describe the spin, but not the spin of the electron. Dirac recognized that these two components describe the spin of the antiparticle of the electron, which was later called the positron. It has a positive charge, and it has exactly the same mass as the electron.

When Dirac formulated his equation in 1928, he was not sure whether antiparticles really exist. For some time he thought that the proton was the antiparticle of the electron. But in this case there is the problem that the proton mass is not equal to the electron mass. Only in 1931 did Dirac conclude that the antiparticles must exist.

Carl David Anderson, a physicist at Caltech in Pasadena, carried out experiments with cosmic rays. In 1932, in a cloud chamber, he found the trace of a particle which looked like the trace of an electron, but it was deflected in the magnetic field in the opposite direction to an electron, like an electron with a positive charge. A new particle with the mass of the electron, but with positive electric charge, was discovered. This discovery validated Dirac's prediction.

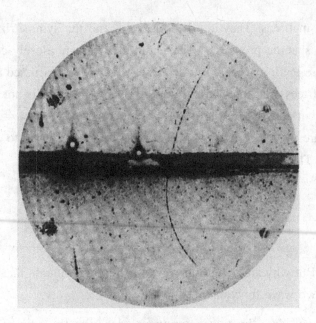

The trace of a positron in Anderson's cloud chamber.

Carl Anderson

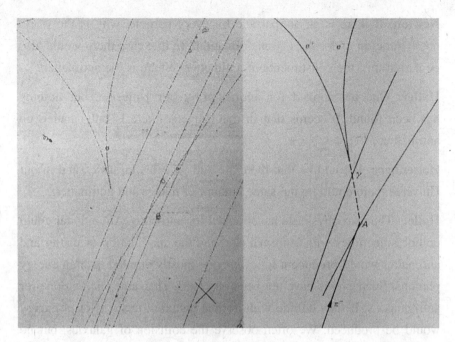

On the left is a bubble chamber picture of the creation of an electron and a positron, on the right is the theoretical interpretation. In a collision a photon is emitted, which after a very short time produces a positron and an electron.

Haller: Anderson, who received the Nobel Prize in 1936, did not know Dirac's prediction. Today we know that every particle has its antiparticle, for example the proton has the antiproton, which has a negative electric charge. In some cases the antiparticle is the same as the particle, for example, the antiparticle of the photon is again the photon. The symmetry between particle and antiparticle is called C. It is one of the fundamental symmetries of nature. But it is not an exact symmetry. The weak interactions, which we shall discuss later, violate the C symmetry.

The antiproton was discovered in 1955 in Berkeley, where an accelerator which could accelerate protons to an energy which was high enough to produce the antiproton had been built. Later the antineutron was also discovered.

Newton: If there is an antiparticle for every particle, why is the world not symmetric under the C transformation? In that case there would also be anti-atoms, the constituents of antimatter. Where is the antimatter?

Haller: One has looked for antimatter in our Universe, but nothing has been found. It seems that in our Universe there is only matter, no antimatter.

Heisenberg: It could be that there are anti-galaxies far away, and in our Universe there could be the same amount of matter and antimatter.

Haller: The astrophysicists have looked for anti-stars. An anti-star could collide sometimes with a normal star, and the annihilation of matter and antimatter would produce a lot of energy, mostly emitted as high energy photons. Such events have not been observed. The same can be done for anti-galaxies. If they collide with normal galaxies, again a lot of energy would be produced. We often observe the collision of galaxies, but no energy is emitted. Today we know that antimatter does not exist in large quantities anywhere in our Universe.

Newton: But then I have a problem. What about the C symmetry? Suppose that our universe was created in a Big Bang. In this event matter and antimatter should have been produced in equal proportion. Thus in the Universe there should be both matter and antimatter.

Haller: Particle physicists have found a solution to this problem. They assume that the symmetry between particles and antiparticles is broken slightly. Right after the Big Bang, the Universe had matter and anti-matter in equal proportion. But the small symmetry breaking becomes relevant. It implies that for about 10 billion pairs of particles and anti-particles one additional particle exists, produced by the symmetry break-ing. But then the particles and antiparticles annihilate. The additional particles do not find an antiparticle to annihilate, and they later form atoms and molecules, which is the matter we have in the Universe today. Thus the very small symmetry breaking leads finally to a big violation of the particle–antiparticle symmetry. Whether this explanation is correct, we do not know.

Feynman: Let us return to the Dirac equation. It is a relativistic form of the Schrödinger equation. Quantum mechanics, combined with Einstein's theory, leads to the Dirac equation and also to the spin.

Haller: I would like to add that in particle physics all particles have some spin. There are scalar particles with spin zero, particles with spin $\frac{1}{2}$, with spin 1, with spin $\frac{3}{2}$ etc.

Newton: I do not know particles with spin $\frac{3}{2}$. What are these particles?

Haller: Yes, they exist, but not as stable particles. A particle with spin $\frac{3}{2}$ is, for example, the famous hyperon Ω^-, a particle with a mass of about 1.8 times the proton mass. It decays, for example, into a proton and two particles with spin zero, which are called mesons. Those are also unstable and decay at the end into an electron or a positron and other neutral particles with spin $\frac{1}{2}$, which are called neutrinos.

Newton: These neutrinos — are they also unstable?

Haller: No, neutrinos are stable like the electrons. They were introduced 1930 by Wolfgang Pauli. At this time there was a problem with the nuclear beta decay of atomic nuclei. Some nuclei decay by emitting an electron. It is easy to determine the mass of the nucleus before the decay and to measure the energies of the electron and of the nucleus after decay. Using Einstein's equation, one was able to check whether the energy is conserved. However, less energy was measured. The energy of the electron was smaller than expected and varied continuously, contrary to expectations. Niels Bohr proposed to give up energy conservation in the beta decay.

Newton: This sounds crazy to me. There must be another way to understand the effect.

Haller: Yes, this way was found by Pauli. He proposed that in the decay another neutral particle is emitted, which was later called a neutrino. Pauli was not happy with his neutrino hypothesis, since he thought that he had invented a particle which could never be seen. However, he was wrong.

If a neutrino collides with a nucleus, the nucleus is often changed to another nucleus, and the neutrino changes to an electron. This reaction was observed in 1955 in an experiment, done by Frederick Reines and Clyde Cowan, using the neutrinos emitted from the Savannah River reactor in South Carolina. This was a very big reactor, producing the nuclear material used later in atomic bombs. Thus it took 25 years to observe the neutrinos after they were introduced by Pauli.

Newton: The neutrinos are peculiar ghost particles. But the energy conservation was saved by Pauli — a good idea. You mentioned the unstable mesons, which have spin zero. Thus all the stable particles in our Universe have spin.

Haller: Yes, that is true, with one exception, which I will mention later. The particles with half-integer spin, i.e. spin $\frac{1}{2}$, spin $\frac{3}{2}$ etc., are called fermions, named after Enrico Fermi, an Italian physicist, who won the Nobel Prize in 1937 and moved to the United States in the same year. The particles with integer spin, i.e. spin 0, spin 1, etc. are called bosons. There are bosons which take part in the strong interactions — those are called mesons. The particles with half-integer spin, which take part in the strong interactions, are called baryons. Thus the protons and neutrons are baryons.

The stable particles in our universe are the proton, the electron, the neutrino — they all have spin $\frac{1}{2}$. All other particles are unstable, with the exception of the photon. A photon cannot decay since it is massless. An electron cannot decay since there is no particle which is charged and has a mass less than the electron mass.

Einstein: You mentioned that the proton is stable. Perhaps this is not true. It could, for example, decay into a positron and a photon.

Haller: Yes, particle physicists have developed theories which predict the proton decay. Big detectors have been built. The largest one, called Kamiokande, is south of Toyama in Japan, near the small village of Kamioka. But no decay has been found so far. The present limit on the lifetime of the proton is 10^{32} years.

Newton: Sorry, I do not understand this. I read recently that the astrophysicists think our Universe has existed for about 14 billion years. This

The Kamiokande detector in Japan.

would also be a limit on the proton lifetime. But you mentioned a much larger limit.

Feynman: Mr. Newton, do not forget that the proton is a quantum mechanical system. What counts here are the probabilities. A neutron, for example, has a lifetime of about 15 minutes. This implies that, on average, neutrons live 15 minutes. But there are also neutrons, which decay faster, i.e. after one minute. In Kamioka one is searching for those protons which decay fast. In the experiment one does not observe a single proton, but very many of them, about 10^{31}. The detector is a big pool of purified water, which is investigated with about 11,000 photo-multipliers. These instruments can see the photons which are emitted in the proton decay.

Haller: The big Kamiokande detector was built in order to test specific theories which predict the proton decay. According to those theories, the protons have a lifetime between 10^{31} and 10^{34} years. The lower part of this range has already been excluded by experiments.

Newton: If they find a decaying proton, it would be an important discovery. I would like to know more about the theories which predict the proton decay.

Einstein: Ok, but now it is lunch time. I propose that we take a walk through the forest and have a picnic. I shall go to the kitchen and pack the things we need.

ELECTRONS AND PHOTONS

10

After some time the four physicists returned to the garden of Einstein's house.

Heisenberg: Today we shall discuss the development of quantum mechanics related to the work of Wolfgang Pauli and myself shortly after 1930. We started with Dirac's equation and tried to understand the interaction of electrons and photons. Finally we invented the theory of quantum electrodynamics, often called QED. You, Mr. Feynman, and Julian Schwinger subsequently developed this theory to a much greater extent.

To start with, I would like to introduce a new particle which is used quite often in QED: the virtual particle. My uncertainty relation describes the uncertainty of space and momentum. But there is also an uncertainty between energy and time. This relation looks like:

$$(\Delta E)(\Delta t) \approx h.$$

It implies that in a very short time one can borrow a lot of energy: a particle can turn into another particle very quickly. For example, a photon can become an electron–positron-pair, and after a short time the pair again becomes a photon. Particles which do not exist as free particles, but can appear in short time intervals, are called virtual particles.

The Dirac equation is a relativistic equation, like the Maxwell equations. Pauli and I tried to find an interaction which is also consistent with Einstein's theory. The simplest interaction is that of the electrons and positrons with the potential of the electromagnetic field. The electric and magnetic field strength can be derived from a potential. This potential is a so-called four-vector, which is a normal vector plus a fourth component, which is related to the time. A four-vector has four components. These four functions determine the six functions which describe the electric and magnetic fields.

Einstein: One could also consider a direct interaction of the electron and the electromagnetic field strength. The potential is a strange construction which cannot be measured, unlike the field strength. Why did you and Pauli consider the interaction with the potential?

Heisenberg: Well, the interaction with the potential was simpler. But I would not worry about the fact that the potential cannot be measured. The Dirac field, usually described by the spinor ψ, cannot be measured either. We could also have considered a direct interaction between the field strength and the spinor ψ, but we tried to construct a simple interaction. Our interaction turned out to be very successful, at least for the electrons, but not for the protons.

Around 1930, the experimentalists measured the magnetic moment of the proton. Pauli used to joke about these experiments. He said that the proton is a Dirac particle, thus its magnetic moment is fixed by the Dirac equation, like it is for the electron. The neutron has no electric charge, and Pauli predicted that the magnetic moment of the neutron should also vanish.

Feynman: Pauli was completely wrong. Today we know the magnetic moments of the proton and neutron exactly. The magnetic moment of the proton is $\mu = 2.793 \; \mu_N$. Here μ_N is the magnetic moment of a Dirac particle with the mass, given by the proton mass. Thus the magnetic moment of a proton is almost three times as big as the magnetic moment of a Dirac particle. Furthermore, the magnetic moment of the neutron is different from zero: $\mu = -1.9131 \; \mu_N$.

Newton: Protons and neutrons are not elementary, but consist of quarks, and thus it is evident that they are not elementary Dirac particles. Their magnetic moments are complicated quantities; nobody can calculate them. But I think that the quarks should be elementary Dirac particles.

Haller: Yes, I agree. Protons and neutrons consist of quarks and have an internal structure, unlike the electron. The magnetic moments depend on details of the wave functions, and nobody can calculate them. However, the quarks are elementary Dirac particles.

Let me mention something else. Vladimir Fock in Russia investigated in 1926 the electromagnetic interaction of a complex scalar field. He found that this interaction has an interesting symmetry provided by the phases of the fields. If we multiply a complex field with a complex number, the basic equations are not changed. Such a phase rotation is an element of a mathematical group, a U(1) group.

Einstein: But I have to multiply the field here and far away, on the moon, by the same phase. This is rather strange. I believe that the Dirac field on the moon should be independent of the field here on earth.

Vladimir Fock, Professor of Physics in Leningrad, Soviet Union

Heisenberg: Fock knew this, and he used a particular symmetry. He multiplied the field everywhere by the phase $e^{i\lambda}$, where the phase parameter λ depends on space and time. In this case the equation of motion is not valid, since new terms involving the derivative of λ appear in the equation. Fock's idea was to change the electromagnetic potential by a term given by the derivative of λ. Thus both the field and the electromagnetic field are changed — the change of the field is compensated by the change of the potential. Such a symmetry has also been found later by Hermann Weyl in Germany. He called the invariance of the theory under a change of the phase "gauge invariance". A theory with this symmetry is called a gauge theory.

Feynman: Yes, and this interplay between the phase of the Dirac field and the potential is only possible if there is an interaction. Without an interaction the symmetry is violated, thus the theory with the electromagnetic interaction has this new gauge symmetry. You and Pauli considered the right theory — the interaction is not with the field strength, but with the potential.

Haller: Today we know that this theory is the right one. In the gauge theory, the quantum theory and the relativity theory are united and the fields are quantized. The gauge theory is called quantum electrodynamics, QED.

Newton: Is the quantized theory consistent?

Heisenberg: Pauli and I assumed that this was the case, but soon problems arose. We applied the so-called perturbation theory. This method is often used in quantum mechanics. One considers the interaction as a perturbation for the free motion and uses an expansion.

Let me mention a simple example in mathematics. The function $1/(1 - x)$ can be written as a Taylor expansion:

$$1/(1 - x) = 1 + x + x^2 + x^3 + \cdots .$$

If x is small, the expansion works quite well; one just needs the first two or three terms. Of course, one can also calculate this function directly. However, in QED this is not possible — one can only use the perturbation expansion. But it turns out that the perturbation theory is *not* consistent. This was realized in the early '50s, especially by you, Mr. Feynman. You were working at that time at Cornell University in Ithaca, but in 1950 you moved to the California Institute of Technology in Pasadena.

Feynman: Yes, while at Cornell I calculated the effect of the electromagnetic interaction for the electron, especially for the mass of the electron. An electron emits a photon, and shortly afterwards it captures the photon again. This interaction can be calculated, but I obtained a senseless result for the mass of the electron.

Newton: Why senseless? If you can really calculate the mass of the electron — this would be a real discovery.

Feynman: Yes, but it turned out that the mass of the electron is infinite.

Newton: In that case the theory does not make sense.

Feynman: No, but I had an interesting idea. The mass of an electron without the electromagnetic interaction is a quantity which nobody can observe. We see only the physical electron with a mass of about 0.5 MeV. I absorbed the infinity into the mass of the unphysical electron without the interaction. When the calculations are done, one has infinities, but at the end they disappear. This process is called renormalization. Of course, the mass of the physical electron cannot be calculated. It remains a parameter which has to be taken from experiment. Dirac did not like this idea. Whenever I met him, he discussed this problem with me, but he could not find a solution.

The renormalization program has had a number of successes. Julian Schwinger, for example, calculated the magnetic moment of the electron, and the experimentalists measured this magnetic moment very precisely. Previously Dirac had predicted the magnetic moment. It was simply given by $(e/m) \times$ spin, where m is the mass of the electron. The magnetic moment is given normally by the g factor. According to Dirac this factor should simply be 2. Today we know this factor very precisely: $g = 2.0023193043718$.

Newton: Thus this factor is not exactly equal to 2, but slightly larger.

Feynman: Yes — and this departure from 2 had been calculated by Schwinger. In 1912, Arnold Sommerfeld at the University in Munich found that the strength of the interaction between the photons and the electrons can be described by a number without a physical dimension, which he called the fine structure constant, usually called α. He called it the fine structure constant because it is relevant for the fine structure of the atomic energy levels. The constant is given by the electric charge e, Planck's constant h and the velocity of light c: $\alpha = 4\pi e^2/hc$. Thus, in this constant, the theory of electrodynamics, the theory of relativity and the quantum theory come together. Using the experimental values, one obtains $\alpha \sim 1/137$. Today the fine structure constant can be measured with very high precision using lasers. The present value is: $\alpha = 1/137.035999$.

Arnold Sommerfeld (1868–1951)

The number 137 has fascinated many physicists since 1912. Wolfgang Pauli tried throughout his life to calculate this number, without success. When he died in 1958 in a hospital in Zürich, it was discovered that he died in the room 137. This might have been a coincident, we do not know.

Schwinger calculated the departure of g from the Dirac value in terms of α. He found $g = 2(1 + \alpha/2\pi)$. Schwinger's value agreed completely with the experimental value — a big success for QED.

Haller: Yes, this was a big success for the new quantum field theory of QED. Schwinger calculated the correction which is given by a virtual photon emitted before the measurement of the magnetic moment, and afterwards again absorbed. It can be described very well by a diagram which you, Mr. Feynman, have introduced, and which is now called a Feynman diagram. Schwinger's correction is then given by the following diagram:

The Feynman diagram for the magnetic moment of the electron (above without any correction, below the first-order correction calculated by Schwinger).

Newton: Schwinger calculated this diagram?

Feynman: No, Schwinger did not know my diagrams at the time. Even later he did not use them, since he did not like the diagrams. Thus his calculations were much more complicated.

Schwinger presented his result about the magnetic moment at a small meeting in 1947 on Shelter Island near New York.

It was a long calculation, and took about one hour. I was at this meeting and repeated his calculation at night in my hotel room, using my diagrams. I obtained the same result, but in only a few minutes. At this moment I realized that I had found something important, and I published it in Physical Review.

A discussion at the Shelter Island Meeting — Feynman is in the middle, Julian Schwinger is on the right.

Haller: Indeed the Feynman diagrams are very useful. Without them, the progress in quantum field theory would have been much slower. Using the diagrams, one can calculate many processes quite easily. For example, the magnetic moment of the electron receives contributions from higher orders, e.g. the exchange of two virtual photons, which is of second order. It is possible to write the result in a power series of α:

$$\alpha/2\pi + B\alpha^2 + C\alpha^3 + D\alpha^4 + \cdots.$$

The numbers B, C etc. have to be calculated.

Einstein: The magnetic moment of the electron has been measured with high accuracy. Does the result agree with the theoretical calculations?

Haller: The theory works very well. The agreement between experiment and theory is better than 1 to 10 million, thus the theory of QED seems to be correct.

Newton: QED — this theory is obtained by combining quantum mechanics and Einstein's theory of relativity. Mr. Heisenberg, you and Mr. Pauli and later Mr. Feynman have created a wonderful theory.

Haller: The theory of QED is not an isolated theory — today we know that it is part of a larger theory, which includes also the weak interactions. But right now we shall not speak about this.

Newton: QED has two free parameters, the mass of the electron and the fine structure constant. Can one say something about these constants? The fine structure constant in particular could be calculated.

Heisenberg: The electron mass has a dimension, and it cannot be calculated. Many physicists tried to calculate the fine structure constant. Wolfgang Pauli made many attempts, without any success. He died at the early age of 58 in a hospital in Zürich. His previous assistant, Victor Weisskopf, visited him shortly before he died, and he noticed that Pauli was in room number 137; we do not know whether this was by coincident or deliberate. Sommerfeld's number 137 is still mysterious — nobody has an idea how to calculate this number.

Haller: My friend Murray Gell-Mann believes that the numerical value of the fine structure constant is a cosmic accident. Our world was created in the Big Bang about 14 billion years ago. In the beginning the world consisted of a hot mixture of electrons and quarks, and the fine structure constant fluctuated. Then the universe expanded and became cool, and the fine structure constant was frozen accidentally to about the inverse of 137.

Einstein: If the Big Bang were repeated, would the fine structure constant be different?

Haller: Yes, the fine structure constant is just a cosmic accident.

Feynman: I read once in a book that our world would look quite different if α had a different value, such as 1/140.

Haller: Yes, if α were different, some macromolecules which are relevant for our life would not exist, and nor would we.

Einstein: This is strange — our life and the numerical value of the fine structure constant are then strongly correlated.

Haller: Gell-Mann and many other physicists believe that there was not only one Big Bang, but there were many of them, perhaps infinitely many. Our world should not be called a Universe, but a multiverse.

Every Big Bang led to another fine structure constant, and in our world it is accidentally close to the inverse of 137. Only in this Universe is there life, and thus the connection between life and the value of the fine structure constant is seen, but it is a pure accident.

Einstein: Again an accident, as everywhere in quantum theory. No, I do not believe that the value of the fine structure constant is an accidental value — some day one should be able to calculate this constant.

Newton: Mr. Einstein, if the value of the fine structure constant is an accident, it does not make sense to calculate its value. It is a cosmic accident. Nobody can calculate this number.

Einstein: I must confess that this physics is taking a direction which I do not like. Today I would not become a physicist, but the attendant of a lighthouse on an isolated island.

Haller: Now it is time for dinner. I propose that we again go to Potsdam.

A few minutes later they were in a taxi and drove to the restaurant "Alte Post".

COLORED QUARKS
AND GLUONS

11

The next morning, the physicists met once more on the terrace of Einstein's house.

Einstein: Atoms can be described very well by the quantum field theory of QED, which is a theory of electromagnetism, of quantum mechanics and relativity theory. We know that the protons and neutrons are not elementary, but bound states of quarks. For atomic physics this is not relevant. But I would like to understand the nucleons and the nuclear forces. Is there a consistent theory of the nuclear forces?

Feynman: Yes, but before we come to this theory, we need to understand some phenomena of particle physics. The physics of elementary particles started around 1950. When cosmic rays were being investigated, many new particles were discovered; first the π-mesons, then the K-mesons, then new heavy particles, the hyperons. New symmetries were introduced, in particular the symmetry of SU(3), proposed by Murray Gell-Mann at Caltech and by Yuval Ne'eman in Israel.

In 1964, Gell-Mann and George Zweig proposed that the nucleons were bound states of elementary particles, which Gell-Mann called quarks. Gell-Mann was able to explain the SU(3) symmetry using quarks. Not many believed that this was the right way, but today we know that Gell-Mann and Zweig were in fact correct.

Since the quarks have strange electric charges, Gell-Mann did not send his paper to *Phys. Rev. Lett.*, and he had a feeling that the referees would reject it. He sent it instead to *Phys. Lett.*, where it was printed, and this short paper turned out to be the most successful one ever printed by that journal.

George Zweig was a Ph.D. student of Gell-Mann. After his graduation he went to CERN, where he wrote a long paper about the quarks, which he called "aces". At that time, physicists at CERN had to publish their papers in European journals. Zweig intended to publish his results in *Phys. Rev.*, but was denied the permission to do so, and thus he never published his article. Even today, Zweig's article can only be read as a CERN preprint; he should have published his ideas in *Nucl. Phys.*

Newton: Why do the quarks have these strange electric charges?

Feynman: This can easily be seen. For the protons and neutrons one needs two quarks. Gell-Mann named them "u" and "d", from "up" and "down". The proton has the structure (uud), and the neutron has the structure (ddu). The charges can then be easily calculated: $2Q(u) + Q(d) = +1$, $Q(u) + 2 Q(d) = 0$, which gives $Q(u) = 2/3$ and $Q(d) = -1/3$. Thus the charges are not given by integral numbers.

Haller: The charges of the quarks are indeed rather strange. For this reason many physicists did not like the quark model. But today we know that the quarks exist inside the nucleons. Their charges were measured with electron beams, and indeed they were 2/3 and −1/3. The experiments

The Stanford Linear Accelerator Center (SLAC).

were carried out at the Stanford Linear Accelerator Center (SLAC) in Stanford, California.

Eight years later, in 1972, Gell-Mann and his young German collaborator Harald Fritzsch discussed for the first time the possibility that a simple field theory, similar to QED, is able to describe the phenomena of the strong interactions. Today this theory, called quantum chromodynamics (QCD for short), is generally accepted as the theory of the strong interactions. It is in agreement with the experiments. The theory of QCD is a unification of the relativity theory, quantum mechanics and a new color theory of the quarks, which we shall discuss later.

Newton: Do the quarks have a particular color?

Feynman: The color is not a real color, just a particular index. But this index solved a serious problem. In the simple quark model of Gell-Mann and Zweig, there are four excited states of the nucleon, which are unstable and have spin 3/2. They were discovered in Berkeley and are called the delta particles. Let us consider the Δ^{++}-particle, which has a mass of about 1230 MeV and the electric charge +2. In the quark model this particle consists of three u-quarks, and its electric charge is three

Murray Gell-Mann (right) and Harald Fritzsch in Berlin (1995).

times 2/3: 2. The wave function of this particle is very simple. The three *u*-quarks are in an *s*-wave, and their spins point to the same direction. The result is a particle with spin 3/2 and the structure (*uuu*).

Heisenberg: But then there is a problem — this particle consists only of *u*-quarks, and the wave function is totally symmetric, not anti-symmetric, as required by the Pauli principle, since the quarks are fermions.

Feynman: Yes, the quarks should obey the Pauli principle; the wave function should be antisymmetric, if two *u*-quarks are interchanged. However, in the quark model the wave function is symmetric. For this reason the quark model was not accepted by many physicists. In 1971, Fritzsch and Gell-Mann found the solution. They assumed that besides their electric charges, their masses and their spins, the quarks have another property, some kind of threefold charge. Fritzsch and Gell-Mann called this charge the color of the quark — there are red quarks R, green quarks G and blue quarks B. But this color is not a real color, of course.

Let us consider the wave functions of the baryons, taking into account the new color quantum number. For example, let us take the wave function, describing three red quarks: (RRR). Again we have the problem with the Pauli principle, since this state is again symmetric under the exchange of two quarks. But there is a wave function, which is antisymmetric, the state (RGB − RBG + BRG − BGR + GBR − GRB).

Newton: What determines the number of colors? Can we also take two colors, or perhaps four?

Haller: No, there must be three colors, since the proton consists of three quarks. If there were only two colors, the proton would consist of two quarks. In 1964, Gell-Mann was wondering why the proton consists of three quarks, not two quarks or four quarks. The color quantum number gives the explanation — three colors imply that the proton is composed of three quarks. The three colors of the quarks led to a new symmetry, the color group, which is described by the mathematical group SU(3).

Three colors of the quarks are also needed to describe the electromagnetic decay of the neutral π-meson into two photons. The decay rate depends on the number of colors. If there are no colors, the decay rate is about 1/9 of the actual decay rate. This was one of the arguments against the quark model. But for three colors the decay rate is $3 \times 3 = 9$ times larger and agrees perfectly with the experiment.

Haller: The bound states of quarks, which appear as physical states, are singlets of the color group SU(3). For those states the colors of the quarks neutralize, and sometimes one speaks of white states.

The rotation of the colors describes a symmetry which is similar to the phase rotation in QED. Fritzsch had worked on gravity in East Berlin, and as a preparation he had studied in detail the Yang–Mills field theory.

Einstein: I am an expert on the theory of gravitation, but I have no idea what a Yang–Mills theory is.

Feynman: Let me explain — in 1954 the American physicist Chen Ning Yang and his collaborator Robert Mills worked at the Institute for

Advanced Study in Princeton, the institute you were also visited at that time. Yang and Mills were interested in the strong interactions and tried to understand the details of the isospin symmetry. They had the idea to consider the isospin symmetry like the gauge symmetry in electrodynamics. This implied the existence of new force particles, analogous to the photon in electrodynamics. Since the isospin group has three charge operators, there would have to exist three new force particles with spin one. These force particles interacted with the protons and neutrons.

Even before 1954, a theory of this kind had been studied by Wolfgang Pauli but Pauli did not publish his work. He wrote his results down in a letter, which he sent to a colleague at the Institute for Advanced Study in Princeton. It is likely that Yang has seen the letter, but he did not mention the letter in his publication.

But neither Pauli nor Yang and Mills were able to introduce a mass for the force particles. Thus the model could not be realistic, since massless particles describing the interactions of the protons and neutrons, did not exist. Nevertheless, Yang and Mills published their results, and today we call this theory a Yang–Mills field theory.

Fritzsch knew much about the Yang–Mills theory, and one day he discussed with Gell-Mann the possibility of using a Yang–Mills theory in the color space. Gell-Mann did not like this proposal at the beginning, but after a short time he realized that this might be a very good idea. The problem with the massless force particles did not exist in such a theory, since all colored objects, including the force particles, would not exist as free particles, but they would be permanently bound.

In 1972, Fritzsch and Gell-Mann published their new theory of the strong interactions. Since this theory was very similar to quantum electrodynamics (QED), they called this theory quantum chromodynamics (QCD). In this theory, the interactions between the quarks were provided by eight force particles, eight gauge bosons.

At the beginning many physicists did not take this theory seriously, but it has turned out that the QCD-theory is able to describe the dynamics of the quarks inside the atomic nuclei. Today it is the theory of the strong interactions. All experimental results are in agreement with the predictions of QCD.

Newton: I guess that the color property of the quarks provides a reason why the quarks do not appear as free particles.

Haller: Yes, one requires that physical particles be color singlets. With three quarks one can easily form a color singlet. One has to take a red, a green and a blue quark. Two quarks cannot be a color singlet, however. But a quark and an antiquark can also form a color singlet, namely the sum "antired ~ red + antigreen ~ green + antiblue ~ blue". Those states are the mesons, for example the π-mesons. The positively charged π-meson has the structure "anti-d-quark + u-quark".

I would like to mention the main difference between the theory of electromagnetism, QED, and the color theory of the quarks, QCD. The gauge symmetry of QED is very simple. It is given by phase rotations of the electron field. Such a rotation can be described by a single parameter. The mathematicians call such a symmetry an Abelian symmetry, referring to the Norwegian mathematician Niels Henrik Abel.

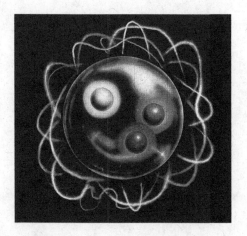

A proton consists of three colored quarks.

The symmetry of the color theory is more complicated. This can be seen if we consider an example from geometry. If there are two dimensions, the rotations describe an Abelian symmetry. If we add one more dimension, then we have three different rotations, around the x-axis, around the y-axis and around the z-axis. An arbitrary rotation can be described by three parameters. The symmetry is a non-Abelian symmetry. A gauge theory which has a non-Abelian gauge symmetry, is called a non-Abelian gauge theory, or a Yang–Mills theory.

In QCD there are eight gauge bosons, which Gell-Mann called "gluons", since the quarks are "glued" together inside a hadron.

Einstein: I do not like the name "gluon" — English and Greek elements in one word. I would have preferred a better name, chromon for example, since "chromos" is the Greek word for color.

Haller: Fritzsch did not like this name either. He indeed proposed the name "chromon". But Gell-Mann did not like it, and we are now stuck with the gluons. However, Fritzsch and Gell-Mann introduced an interesting name for the theory: quantum chromodynamics.

Einstein: Yes, this is a very good name. Quantum chromodynamics sounds even better than quantum electrodynamics. Is the theory as good as its name?

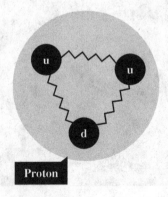

Three quarks, bound by gluons, forming a nucleon.

Haller: Yes, the theory describes the strong interactions very well. All experiments are in agreement with the theoretical predictions, based on QCD. But most of the theoreticians found that the theory is too complicated; especially they did not believe that the color symmetry is an exact symmetry of nature.

Einstein: But this feature of QCD I find very interesting. A symmetry which is not broken — this is something beautiful. Because the symmetry is unbroken, it cannot be seen directly. I like this — I like the theory of QCD.

Feynman: Even before the theory of QCD was introduced, I had the idea that the quarks could be seen indirectly in the electron–positron annihilation at high energies. When a quark is produced, it will be confined. But it has a high momentum, and this momentum splits into many hadrons, mostly mesons. These particles form a jet of hadrons. However, the sum of the momenta of these hadrons should be equal to the quark momentum. Thus in the electron–positron annihilation one should see two jets of hadrons. These quark jets were discovered at DESY in 1979.

Then the experimentalists discovered that gluons should also exist. One expects that after the annihilation of an electron and a positron, a quark, an antiquark and a gluon should sometimes be produced. Thus one should see three jets, and they were indeed observed at DESY in 1979.

Measuring these jet phenomena, one was able to find out something specific about the strength of the interaction between the quarks and gluons. This strength is described by a parameter, which is usually denoted by α_s — it is the strong interaction analogue of the fine structure constant α in QED. With the LEP accelerator at CERN one determined this value to be $\alpha_s = 0.12$.

A non-Abelian gauge theory like QCD has a structure which is quite different from the structure of an Abelian gauge theory like QED. If an electron and a photon interact, the electron and the photon change their momentum, nothing else. In the interaction of a quark with one of

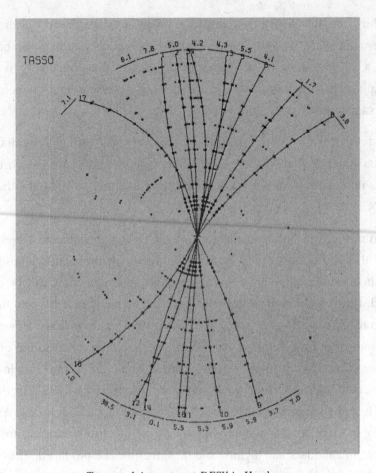

Two quark jets, seen at DESY in Hamburg.

the eight gluons, in general, the color state of the quark changes. For example, a red quark changes to a green quark.

Let me write down a little dictionary for QED and QCD:

QED	QCD
electron, muon	quarks
electric charge	colour charges
photon	gluons
atom	nucleon

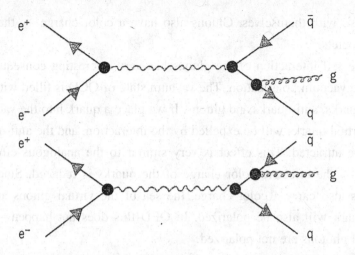

The production of a quark and an antiquark in the electron–positron annihilation — the quark or the antiquark emits a gluon.

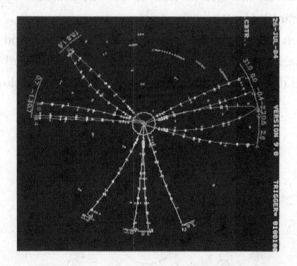

Three jets, seen by the TASSO detector at DESY.

The photons are electrically neutral, which in particular means that they cannot interact directly with themselves. Millions of photons can fly together through space in a laser beam, without disturbing each other. Gluons cannot do this, since they interact not only with the quarks, but also with themselves. Gluons also have a color charge — they are color octets.

The self-interaction of the gluons has very interesting consequences for the vacuum polarization. The vacuum state of QCD is filled with virtual quarks, anti-quarks and gluons. If we place a quark into the vacuum, the virtual quarks will be expelled by the interaction, and the anti-quarks will be attracted. This effect is very similar to the analogous effect in QED — the effective color charge of the quarks is reduced. Since the gluons also carry a color charge, the sea of the virtual gluons around the quark will also be polarized. In QED this does not happen — the virtual photons are not polarized.

At the end of the '60s, the effects of the vacuum polarisation in the non-Abelian gauge theories were investigated. The first calculations were done by the Russian theorist Iosif Kriplovich and later by the Dutch theorist Gerard 't Hooft.

More detailed calculations were made in 1973 by David Gross and Frank Wilczek in Princeton, and by David Politzer at Harvard University. They found that in QCD the virtual gluons increase the color charge of

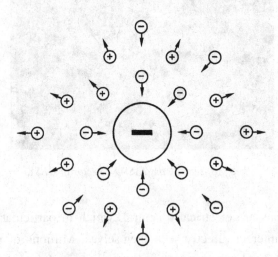

The screening of the electric charge of an electron by the virtual electron–positron pairs.

Gerard 't Hooft

a quark, unlike in QED. This implies that the strength of the interaction decreases at small distances, and increases at long distances. This effect was called asymptotic freedom.

Haller: The fine structure constant of electromagnetism increases at very small distances. In QCD the opposite is true: the gluons contribute to the vacuum polarization, and their contribution is negative. Thus the coupling constant α_s increases at small distances. The experiments give the following value for the coupling constant at the mass of the Z-boson, about 91 GeV: $\alpha_s \approx 0.12$. The coupling constant is smaller than one, and the methods of perturbation theory can be used. But at very small distances the coupling constant increases above one, and perturbation theory fails. This indicates that the quarks and the gluons may not exist as free particles — they are confined. However, a real proof for the confinement is still missing.

Let us consider the interaction between a quark and an anti-quark. At very small distances, i.e. smaller than the typical scale of the strong interactions of about 10^{-13} cm, the force is like the electromagnetic

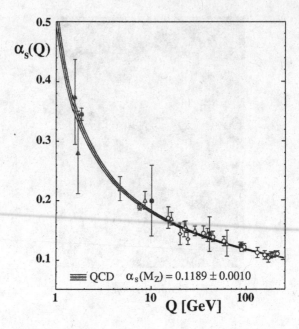

The behavior of the QCD coupling constant as a function of the distance. The coupling constant is not constant, but decreases at small distances.

force, which decreases like the square of the distance. The QCD interaction can be described, like the electric interaction, by field lines. At small distances these field lines are similar, but at large distances the self-interaction of the gluons becomes relevant, and the field lines attract each other. At large distances the field lines become parallel to each other. Thus a quark and an anti-quark are connected by a tube of field lines. The force between quark and anti-quark is constant, and the quarks and anti-quarks appear as confined objects.

At small distances the forces become rather weak. This explains why the quarks appear to be pointlike objects in the scattering of electrons and protons. The disappearance of the strong force at very small distances is called asymptotic freedom.

The colored quarks do not exist as free particles, like any other objects with color. Only color singlets are free particles, like the proton.

The simplest color singlets are bound states of a quark and an anti-quark, the mesons. The first meson, the π-meson, was discovered in 1947 in the cosmic rays. It has a mass of about 140 MeV. The bound states of three quarks, like the proton, are called baryons.

Einstein: The particles you have mentioned so far consist only of quarks. But two gluons, which are octets of color, can also form a color singlet. Such a particle would be neutral, and it would consist only of gluons. Have such particles been found?

Haller: This is a difficult question. Fritzsch and Gell-Mann introduced these particles, which they called glue-mesons. The experimentalists looked for them, but as of today have still had no success. The glue mesons can easily mix with neutral mesons, consisting of quarks. One has studied the neutral mesons, but so far no evidence has been found that these mesons consist at least partly of gluons. The experimentalists are continuing to look for such states.

But now it is time for lunch — let us stop the discussion.

MASSIVE NEUTRINOS

12

Haller: This afternoon I would like to discuss an effect which is connected to the weak interactions, in particular to the neutrinos. But it also has to do with quantum mechanics. As you know, neutrinos were introduced by Wolfgang Pauli in 1930. They are neutral relatives of the electrons – they have no electric charge, therefore they do not interact electromagnetically, but they do have a weak interaction. Both the electrons and the neutrinos are often called leptons, which means something like "light particles".

Today we know that in our world, besides the electron and its neutrino, there are more leptons. There is a charged lepton, the muon, which has a mass of about 200 times the mass of the electron. This particle was discovered in the cosmic rays in 1936. It is unstable and decays into the electron and neutrinos after its creation in a collision in about 10^{-6} seconds. The neutral partner of the muon is also a neutrino. It was found out that this neutrino is different from the electron neutrino — it is called the muon neutrino.

In 1975, another charged unstable lepton was discovered at the Stanford Linear Accelerator Center (SLAC), which is not a light particle like the other leptons, but almost twice as heavy as the proton. It is called the tau lepton or tauon. The neutral partner of the tau lepton is the tau neutrino. With the LEP accelerator at CERN it was found that there are no more neutrinos. Thus in our world there are three neutrinos and three charged leptons.

Newton: Again the number three — in QCD there are three colors, now we have three neutrinos. The number three seems to play a special role in our universe.

Feynman: We have no idea why the number of colors and neutrinos is the same. But the number of leptons seems to be connected with the number of different quarks. The proton and neutron consist of two different quarks, u and d. We like to associate these two quarks with the electron and its neutrino. The two quarks and the two leptons together are called a family. But, besides this family, there are two more. The second family consists of the two heavy quarks s and c, the strange and the charmed quark, the muon and its neutrino. The new quarks are constituents of new heavy particles, which are unstable and decay finally into the proton, the electron and neutrinos.

The tauon and its neutrino are the leptons of a third family, which has two very heavy unstable quarks, the b-quark and the t-quark. The t-quark is rather special. It has a very large mass – it is about as heavy as 185 protons. It was discovered at the Fermi Laboratory near Chicago in 1995.

The number of different neutrinos corresponds to the number of lepton–quark families. In our world there are three families. We do not know whether the number of families is correlated with the number of different colors, but there might be such a connection.

Haller: The neutrinos are the only particles which display a quantum-mechanical effect at macroscopic distances. Neutrinos have very small masses and can oscillate into each other. A muon neutrino can become an electron neutrino and vice versa.

Heisenberg: This is peculiar — neutrinos are normal particles, like electrons, but electrons do not oscillate, why do the neutrinos?

Haller: This has to do with special properties of the weak interactions. In 1957, the neutrino oscillations were first discussed by Bruno Pontecorvo, an Italian physicist, who lived in the Soviet Union.

Let us consider the decays of the charged leptons and the weak interactions of the electron. The weak interactions are mediated by charged intermediate bosons, the W boson. If a boson interacts with the electron, the electron is changed to a neutrino:

$$e^- + W^+ \Rightarrow \nu_e.$$

Analogously we can consider the reaction of the W-boson with a muon or with a tauon:

$$\mu^- + W^+ \Rightarrow \nu_\mu, \qquad \tau^- + W^+ \Rightarrow \nu_r.$$

Nothing special happens if the neutrinos are massless. But if they have a mass, it could be that the neutrino, produced in the interaction of the W-boson with the electron, does not have a fixed mass, but could be a mixture of two or three mass eigenstates. The neutrino produced from an electron could, for example, be a superposition of two mass eigenstates:

$$\nu_e = (\nu_1 + \nu_2)/\sqrt{2}.$$

The muon neutrino would then be an orthogonal superposition, with a minus sign between the two neutrinos. In this case the mixing angle is maximal, 45°. If the muon decays, the emitted neutrino is again a mixture of two mass eigenstates.

Newton: Why did the physicists think about such mixtures? Did the experiments indicate that the neutrinos emitted in the weak decays are not mass eigenstates?

Haller: The hypothesis that such mixtures could exist was proposed by Bruno Pontecorvo in 1957. The main argument for the neutrino mixing came from the quarks, where such mixings were known. If a W-boson interacts with a u-quark, a d-quark is created, but not with a probability of 100%. Sometimes an s-quark is created instead. Thus the emitted quark is a mixture of d and s. The mixing angle is not large, about 13°.

Let us now consider the neutrinos. In a beta decay of a nucleus, an electron neutrino is created. The neutrino propagates through space. If the neutrino has zero mass, the neutrino propagates like a photon, with the speed of light. But if the neutrino is a mixture of two mass eigenstates, with different masses, every mass eigenstate propagates with a velocity which is smaller than the velocity of light. The state with the higher mass propagates more slowly than the state with the smaller mass.

Einstein: I understand — if the state with the smaller mass has arrived at a certain point x, the state with the larger mass is not yet there. Thus something is changing slowly.

Haller: Yes — if at the start the neutrino is the superposition mentioned above: $v_e = (v_1 + v_2)/\sqrt{2}$, then we have at the point x a different superposition, say $v = av_1 + bv_2$, and the numbers a and b depend on the position x.

Heisenberg: Thus the neutrino at the point x is no longer the electron neutrino, but a superposition of electron and muon neutrinos; with a probability p it is an electron neutrino, and with a probability $1 - p$ it is a muon neutrino, when I assume that there are only two neutrinos.

Haller: Correct – such transitions should exist, and were first suggested by Pontecorvo. They are called neutrino oscillations. If the mixing angle is 45°, an electron becomes at a certain distance a muon neutrino, and afterwards it turns again into an electron neutrino, etc.

Newton: When an electron neutrino reacts with a nucleus, it turns into an electron. A muon neutrino turns into a muon. Thus these neutrino oscillations could be observed, e.g. by studying an electron neutrino beam at various distances.

Haller: Yes, such experiments have been carried out since about 1975, but for a long time no effects were found. In that year I visited the laboratory in Grenoble and gave a colloquium on the hypothetical neutrino oscillations. Rudolf Mößbauer was the director of the laboratory at that time. He liked the neutrino oscillations and shortly after my visit started an experiment using the small reactor there. He did not find an effect, but he repeated his experiment later near the big reactor close to Bern in Switzerland. Again he found no effect. Nevertheless, he found rather strict limits on the mass differences of the neutrinos, of the order of 2 eV.

Today we understand why Mößbauer did not find anything. Since about 1999, we have known that neutrinos have a mass and that neutrino oscillations exist. But the mass differences between the neutrinos were found to be very small, of the order of 0.1 eV. We do not know whether the masses of the three neutrinos are so small, or whether the three neutrinos are nearly degenerate in mass, and the mass differences are very small. The oscillations give only information about mass differences, not about the masses directly.

Newton: Thus the three neutrinos are superpositions of three mass eigenstates. What is known about the coefficients, or about the mixing angles?

Haller: Yes, I can tell you, but the measurements are not very precise. The three neutrinos look like this:

$$v_e = 0.83...v_1 + 0.56...v_2$$
$$v_\mu = -0.40...v_1 + 0.59...v_2 + 0.71...v_3,$$
$$v_\tau = 0.40...v_1 - 0.59...v_2 + 0.71...v_3.$$

The neutrinos ν_1, ν_2, ν_3 are the mass eigenstates. Neutrino oscillations give information only about the mass differences, not about the masses. The difference of the squares of the masses of the first and the second neutrino is only 8×10^{-5} eV^2, the difference of the squares of the masses of the second and the third neutrino is about 2.5×10^{-3} eV^2.

Feynman: These mass differences are very small. I understand now why Mößbauer did not find anything. If the neutrino masses are not close to each other, like the charged lepton masses, they must be exceedingly small. I guess it could be that the three neutrinos are nearly degenerate in mass. In that case they might have a mass of the order of 1 eV.

Haller: Yes, this is a possibility. But it is more likely that the neutrino masses are much smaller. For example, if we take the first neutrino mass to be zero, the second mass would be about 0.01 eV, and the third mass would be 0.06 eV. Then it would be very difficult to measure the neutrino masses directly.

Above I have given you the three neutrinos in terms of the mass eigenstates. The electron neutrino is essentially composed of two mass eigenstates, the other neutrinos are composed of the three mass eigenstates.

Newton: If this is the case, I would like to know the oscillation lengths. What do you know about them?

Haller: The $\nu_\mu - \nu_\tau$-oscillations in particular have been studied at the Kamiokande detector in Japan. The muon neutrinos are produced in the collisions of the cosmic rays in the upper atmosphere. In these collisions a lot of π-mesons are produced, and they decay mostly into muons and muon neutrinos. These neutrinos react with the matter in the Kamiokande detector. It was found that there is an oscillation between the muon neutrinos and the τ-neutrinos. The oscillation length depends on the energy, but on average it was of the order of 2000 km. The mixing angle between the muon neutrino and the tau neutrino is quite large, about 45 degrees.

Heisenberg: This is remarkable — the oscillation is a quantum effect. Quantum mechanics is now used for a length of a few thousand km. I never thought that this could be possible.

Haller: Yes, with neutrinos it can be done. The reason is, of course, that the neutrino masses are tiny. New experiments are planned. Soon at Fermilab a neutrino beam will be directed northwards. After about 500 km it will pass a detector, and one will be able to measure the mixing angle between the neutrinos quite well. From CERN a neutrino beam will be sent towards Italy, and it will cross a detector in a cave in the Gran Sasso mountain. Also the neutrinos emitted by nuclear reactors are being investigated in detail. Such experiments are being carried out in China near Hong Kong, in France and in Korea.

Einstein: The small neutrino masses and the large mixing angles are very strange. Do you have any idea why the masses are so small?

Haller: We do not really know. Some physicists think that the neutrino masses are so-called Majorana masses. A few years after Dirac, Ettore Majorana invented a mass term for a neutral spin 1/2 particle, for which particles and antiparticles are the same. Neutrinos might be such Majorana particles, and the physicists are searching for rare decays of atomic nuclei, which could proceed, if neutrinos are such particles. In any case — neutrino physics has become very interesting.

But let us finish now our discussion, it is quite late. I propose that we go to a restaurant in Caputh for dinner.

THE MASSES OF PARTICLES

13

The next morning, the four physicists met for breakfast on the terrace. The weather was fine and cloudless. Einstein proposed that they should go sailing and continue the discussion in the boat. After a short time the boat was ready, and soon they were on the lake. The wind was light, and they sailed slowly towards the East.

Newton: Mr. Feynman, I am confused about something in atomic physics. Atoms consist of nucleons and electrons. Both the nucleons and the electrons have a certain mass. Where do these masses come from? Why are the nucleons about 1840 times heavier than the electrons? In Einstein's theory, mass is some kind of frozen energy, and we can express the masses in energy units, for example in electron volts (eV, keV, MeV, GeV etc.). The mass of the electron is about 0.511 MeV, the mass of the proton is about 940 MeV. One should be able to calculate these masses precisely.

Feynman: Mr. Newton, this is a difficult question — we do not understand where these masses come from. In the Standard Model we have the three masses of the charged leptons, the three masses of the neutrinos and the masses of the six quarks–12 parameters, which we do not understand. In addition there are the masses of the W bosons and of the Z boson.

Haller: An attempt was already màde to introduce the masses of the bosons of the weak interactions 40 years ago. Just giving a mass to these bosons is not possible, since the theory then has infinities which cannot be absorbed in measurable quantities, like in QED. This problem does not arise in QED nor in QCD, since in these theories the gauge bosons, the photons and the gluons are massless. Theorists like the English physicist Peter Higgs proposed to introduce the masses of gauge bosons with the help of a scalar field, which is now called the Higgs field. This scalar field generates a spontaneous breaking of the gauge symmetry, and the masses of the gauge bosons are introduced in this way. This is called the Higgs mechanism.

Peter Higgs

Newton: What is a spontaneous breaking of the symmetry?

Heisenberg: Let me give you a simple example. Consider the water here in the lake. It is homogenous. Now we decrease the temperature. The water freezes, and we obtain ice crystals, like the crystals of snow flakes. These crystals have a sixfold structure — they are not homogenous. Thus the symmetry with respect to translations is broken, and this is called a spontaneous symmetry breaking.

Einstein: I understand this. But why does this kind of symmetry breaking have anything to do with the masses?

Haller: The symmetry breaking is generated in the Higgs mechanism by a scalar field. This field is in interaction with itself and with the gauge bosons. The interaction of the scalar field with itself breaks the symmetry in such a way that the scalar field acquires a nonzero value in the vacuum — this is called a vacuum expectation value.

If this model is right, one can determine this expectation value in terms of the Fermi constant, which describes the strength of the weak interaction, for example the beta decay of the neutron. Using the observed Fermi constant, one finds about 294 GeV for the vacuum expectation value. In order to calculate the masses of the W bosons and the Z boson, one needs to know another parameter, which is usually called the weak mixing angle θ_w and can be measured in the experiments. Using the experimental value for this angle, one finds about 80 GeV for the mass of the W boson and 91 GeV for the mass of the Z boson, in good agreement with the experiments.

If the masses of the gauge bosons are generated by such a mechanism, the theory is — like the theory of quantum electrodynamics — a renormalizable theory, i.e. the infinities can be absorbed. But in such a theory there is no possibility to calculate the masses of the fermions, such as the electron mass. In the Higgs mechanism the fermion masses arise, since the fermions interact with the scalar field, and the corresponding unknown coupling constant determines the fermion mass.

The Higgs field describes a neutral scalar particle called the Higgs particle, which still remains hypothetical today. Searches for this particle have been made with the LEP accelerator at CERN, but nothing was found, except a lower limit for the mass of the Higgs particle of about 115 GeV. Perhaps this particle will be found with the new LHC accelerator at CERN, or perhaps another way to generate the masses of the particles is discovered.

Heisenberg: Can you tell us how the Higgs particle might be discovered?

Haller: Let me mention only the simplest way. Suppose the Higgs particle has a mass of about 200 GeV. It could be produced by the collision of two protons at the LHC, and could decay primarily into two Z bosons. These in turn decay into a muon and its antiparticle. Thus one observes two muons with a rather high energy, and two anti-muons. They can easily be observed.

One can also look for other decays of the Higgs particle. Since the mass is unknown, we do not know what the dominating decay possibility is. If the Higgs particle exists, it probably has a mass in the range between 115 GeV and 500 GeV.

As already mentioned, it is not possible to say anything about the lepton and quark masses. The masses of the fermions could also be the consequence of a substructure of these particles, like the proton masses, which follows from the quark–gluon substructure of the proton. The lepton and quark masses exhibit an interesting spectrum, which extends from the tiny neutrino masses of about 0.1 eV to the large t-quark mass of 173,000 MeV.

At the beginning of the last century, one observed the energy levels of the hydrogen atom, which showed a simple structure, but remained mysterious, until the quantum theory was able to explain the spectrum. The mass spectrum of the leptons and quarks is still not understood. We do not know how long it will take until we have a theory of these masses.

So far we have discussed the masses of the leptons and quarks and the masses of the weak bosons. For the stable matter in the universe only the electron mass is relevant. But the main part of the mass of the visible matter in our universe comes from the masses of the atomic nuclei, and these are determined by the masses of the proton and neutron.

In the theory of quantum chromodynamics, we can calculate the main part of the proton and neutron mass, which is generated by the permanent confinement of the quarks and gluons. The mass of the proton is essentially the field energy of the quarks and gluons inside the proton — this field energy can be calculated. For the first time it has been possible to calculate a mass, and so we have made some progress.

However, the electron mass remains a mystery. Is this mass also some kind of field energy, e.g. the field energy of the constituents of the electron? In that case the electron would have a substructure. But we do not know how small the radius of the electron is; the experiments give a limit of about 10^{-17} cm. Thus in any case the electron must be much smaller than the proton.

Feynman: Let's stop our discussion now. Mr. Einstein, we should interrupt the sailing and go for lunch.

They were at the end of the Schwielow lake, close to the small village of Ferch, where they went to buy some food. After 20 minutes they had bread, milk, butter, wine, cheese, ham and Hungarian salami. In a meadow near the lake they enjoyed their picnic lunch.

THE FUNDAMENTAL CONSTANTS OF NATURE

14

After their picnic, the physicists went for a hike in the nearby forest, returning to the boat about an hour later and resuming their sailing.

Haller: This afternoon I would like to discuss the problem of the fundamental constants of nature.

Newton: What are these constants?

Feynman: The first constant, the constant of gravity, was introduced by you a long time ago. A basic constant is a number, which appears in the natural laws and which we cannot calculate — it has to be determined by experiment. Your constant of gravity is of this type.

Newton: Yes, I know, and the introduction of this constant caused me some headache. I do not like constants which cannot be calculated. But are there other constants besides my constant of gravity?

Feynman: Yes, and unfortunately we now have many of them. In particle physics, there are at least 32 different constants. Some are just integers: the number of space dimensions, which is three in our world, or the number of time dimensions, which is one. Another constant is the number of colors for the quarks, which in QCD is three, and the number of lepton–quark families.

Newton: Why do you mention the number of space dimensions? It is obvious that there are three dimensions of space.

Haller: No, today there are many physicists who consider nine or ten space dimensions, thus at least six additional dimensions, which however are not there at large distances, but come into play at very small distances.

Feynman: Let us not go into the details now. Last century we introduced a theory, which is able to describe the physics of the elementary particles, the Standard Model. It describes the interactions of the matter particles, i.e. the leptons and quarks, which are described by field theories. A mathematical physicist can write down the model in a few lines.

But most particle physicists take it for granted that the Standard Model is not the final theory, but only a good approximation thereof. In this case, one should find departures from the predictions of the Standard Model in the experiments, but thus far nothing has been observed.

The most famous fundamental constant is the fine structure constant, which is called α, and was introduced 1916 by Arnold Sommerfeld in Munich. It is given by the square of the electric charge e, divided by Planck's constant h and the speed of light c:

$$\alpha = \frac{2\pi \cdot e^2}{h \cdot c}.$$

The fine structure constant α was the third fundamental constant, introduced in physics, the first two being your constant of gravity and the mass of the electron, discovered in 1997. But unlike the gravity constant or the electron mass, the fine structure constant is a pure number without any dimension. Thus one might be able to calculate this number in a final theory. But today we can only determine this number in the experiments. The value is now known with high precision: $\alpha = 1/137.03599976$, the reciprocal is rather close to the integer 137, which is a prime number.

This number 137 is the most famous number in science. It is not only important for atomic physics, but also for all sciences and for engineering. The electromagnetic interaction is described by quantum electrodynamics, and in this theory the fine structure constant is a free parameter.

The particular value of α is quite important for our daily life. If α were slightly different, many things would not be the same since the structure of the atoms and of the molecules depends on this value. Life in the universe is only possible if the fine structure constant and the other fundamental constants have specific values. Why do these constants have those specific values? Nobody knows the reason. One possible explanation is that there are infinitely many universes in our world, and every universe has other fundamental constants. We are living in the universe where the constants are suitable for life. In the other universes there is no life, of course.

Haller: When I was at Caltech, we often had lunch in the Athenaeum, and on such an occasion in 1975 we talked about the fine structure constant. You told me that all theoreticians should write on their blackboard: "137 — how little we know." When I came back from that lunch, I went to your office. You were not there, and I wrote in big letters on your blackboard: "137 — how little Feynman knows." I remember that you kept this sentence on your blackboard for quite some time.

Feynman: Yes, I liked it, and it was even true. Also the strong interactions are described by a quantum field theory, the theory of quantum chromodynamics for the interaction between quarks. In this theory there is also a free parameter, an analogue of the fine structure constant, and this number must be determined by the experiments. But unlike the fine structure constant, the constant of the strong interaction is, strictly speaking, not a constant at all, but depends on the energy studied in the particular case. One finds that this number is about 0.12 at the energy of about 100 GeV.

Haller: Yes, thus the stable matter in our universe is described by the gravity constant G, the fine structure constant α, the coupling constant for the strong interactions $\alpha(s)$, the mass of the electron and the masses of the quarks u and d. These are six parameters, which determine the atomic and nuclear physics.

But if we include the unstable particles, the number of constants increases. The unstable particles decay via the weak interactions, whose strength is given by an analogue of the fine structure constant, another free parameter, and by the masses of the W and Z bosons. Also the masses of the other leptons, the muon and the tauon, are free parameters. These particles are unstable. The muon is about 200 times heavier and the tauon is about 3500 times heavier than the electron. Furthermore, there are four unstable heavy quarks, denoted by s, c, b and t. These quarks decay after their production in a particle collision, and in these decays new parameters appear, usually described by three mixing angles and one phase parameter, relevant for the violation of the CP symmetry.

As we know, neutrinos also have masses, and they mix like the quarks. In general there are the three neutrino masses, three mixing angles and three phase parameters, describing CP violation. Thus in the Standard Model there are 28 fundamental constants.

Einstein: Where do these fundamental constants come from?

Haller: Nobody knows. In the experiments we can determine these constants quite well, but we do not understand their values. The natural constants express our lack of knowledge. Introducing numbers which can only be determined by experiments is not satisfactory for a physicist. Are these constants fixed by the laws of Nature, which we do not yet know? Or are these constants just cosmic accidents fixed by the Big Bang? If the Big Bang were to take place again, we would have different constants.

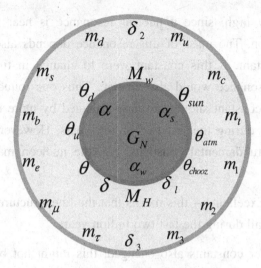

The 28 fundamental constants.

Einstein: 28 fundamental constants — a large number. In my theory of gravity I had only one constant — Newton's constant of gravity.

Haller: There have been attempts to reduce the number of fundamental constants, but so far without any success. But let us consider some constants in more detail, first the fine structure constant. Consider Gabon, a country in the western part of Africa. Two billion years ago there was a big uranium deposit there, not far from the river Oklo. The water of the river penetrated into the uranium. If uranium −235 decays, neutrons are produced. The water acts as a moderator, and the chain reaction starts. Thus in Gabon there was a natural reactor operating for more than 100 million years, two billion years before Enrico Fermi built the first man-made reactor in Chicago.

An exact analysis of the rocks in the Oklo region would give information about the nuclear reactions, which happened two billion years ago, and about the parameters of nuclear physics. Of special interest is the rare earth element samarium. If a samarium nucleus is colliding with a neutron, the neutron might stay in the nucleus, and we obtain a samarium isotope with one neutron more. The cross section for this reaction is exceedingly high, since a nuclear resonance is near the threshold for the reaction. The mass of this resonance depends also on the fine structure constant. If this constant were to change in time, the position of the resonance would also change. Thus one found out that the fine structure constant should not have changed by more than a part in 10^{-16} per year during the last two billion years. However, this is only true if other fundamental constants, e.g. the nucleon mass, remained constant.

Heisenberg: Excellent — this means that the fine structure constant did not change at all during the last two billion years.

Haller: If other constants also changed, this might not be true. Some years ago, astrophysicists studied the fine structure constant by investigating atoms in distant galaxies and distant quasars. Physicists from Australia, England and the United States carried out this experiment at the KECK-telescope in Hawaii. They investigated the atoms of iron, nickel, magnesium, tin and silver, observing about 150 quasars. They discovered a small time variation of the fine structure constant: $\Delta\alpha/\alpha = -(0.54 \pm 0.12) \times 10^{-5}$. They also studied the quasars in various different space directions, but they found no space variation. If one makes a linear approximation, one finds a time change of the fine structure constant of 1.2×10^{-15} per year.

Feynman: But this is in disagreement with the Oklo data.

Haller: Yes, but if other constants also change, e.g. the constant for the strong interaction, the problem disappears.

Einstein: The fine structure constant α is given by $e^2 2\pi/hc$. Thus a time variation of α might come from a time variation of h or of c, or from a time variation of e.

Haller: Mr. Einstein, I am surprised by your remark. If c depends on time, your theory of relativity does not make any sense.

The Keck telescope on the mountain Mauna Kea in Hawaii.

Einstein: Indeed, you are right — for my theory this would be a disaster. Let us forget a time variation of c. But a time variation of h would also be rather bad — quantum theory needs to be changed, and we should forget this. Thus the time variation must come from e.

Heisenberg: You really think that the electric charge e changes in time?

Feynman: Why not? The electric charge describes the force, acting on a charged body, and such a force changes slowly in time, no problem.

Haller: Today we assume that at very high energies the standard model is embedded in a theory of a Grand Unification of all interactions. In such a theory the coupling constants, describing the strengths of the various interactions, come together. At very high energies there is only one unified force.

A quasar, seen by the Hubble space telescope.

A time variation of the fine structure constant can arise only if the constant for the unified interaction depends on the time. But then a time variation of the fine structure constant would imply also a time variation of the weak and strong interactions. A time change of the strong interactions would mean a time change of the masses of the atomic nuclei and of their magnetic moments. I calculated that the strong interaction would change about 40 times faster than the fine structure constant. Then a time change of the magnetic moments of atomic nuclei could be observed in laser physics. Colleagues of mine in Bern, working in laser physics, told me that Theodor Hänsch in Munich might be able to observe such a time change. I had to give a seminar in Munich, and on this occasion I met him.

He told me that he had now finished an experiment, in which he compared the motion of a cesium clock with the hydrogen transitions in his hydrogen spectrometer. I asked him, why this would be interesting. He said: "With the cesium clock I measured a specific hyperfine transition, which depends on the magnetic moment of the nucleus, but the hydrogen transitions are normal atomic transitions, just depending on e. It could be that there is a substantial difference between a hyperfine transition and a normal transition, since the magnetic moment comes into play." Now I became interested.

Heisenberg: I know why you got interested. If the strong interaction changes in time, the magnetic moment also changes.

Haller: Yes, and I would expect that the time change could be observed. Tomorrow a cesium clock has a different frequency than today.

Newton: How big is the effect, which you would expect?

Haller: The effect, seen by the astrophysicists, implies in a linear approximation a change of 1.2×10^{-15} per year. If I multiply this by 40, I get about 5×10^{-14} per year for the change of the masses of the nucleon and for the change of the magnetic moments. At the Max-Planck institute for quantum optics (MPQ) in Munich, Hänsch and his group searched for a time variation of the strong interactions at this order of magnitude, but they did not find any effect. A time change must be less than about 4×10^{-15} per year. The experiment could be improved in the future, up to about 10^{-17} per year.

Einstein: Probably they will find nothing.

Haller: I should mention another experiment. A group from Holland has studied the atomic transitions of hydrogen in very distant quasars, using the European telescope in Chile. The quasars are about 12 billion light-years away. They determined the ratio of the proton mass and the electron mass. It turns out that this ratio is different from the ratio, which

Hänsch's experiment at the MPQ in Munich.

we measure today. The change is of the order of 1/100,000. In a linear approximation one finds that the proton mass changes of the order of 3×10^{-15} per year if the electron mass is assumed to remain constant. This is not much, but big enough that the laser physicists could find an effect.

Feynman: The natural constants are very peculiar. They might be cosmic accidents, but in this case it might be the case, as many cosmologists believe, that our universe is not unique, but is one of many, perhaps infinitely many. Each universe has its own natural constants. One should not speak of a universe any more, but of a multiverse. Of course, if this is the case, there is no way to understand the natural constants in a deeper sense. They are just cosmic accidents.

Einstein: Let us return to the real world, to our universe. Mr. Feynman, how would you ever see another universe? I think these ideas are completely useless. I suggest that we stop sailing now, go back to my house, and then we go out for dinner. Today is our last day, since tomorrow we have to leave.

Newton: Unfortunately our discussions on quantum physics must now come to an end. At the start I had no idea, what a quantum is, now I understand it much better. Quantum mechanics is an impressive theory. I am now a quantum physicist, unlike you, Mr. Einstein. You gave birth to quantum physics, but then you expelled this little baby, and the real physics was done by others, in particular by Mr. Heisenberg. You were a very bad father. But now let us forget the physics and drive to Potsdam to have a good dinner.

THE END

15

Next morning they all went by taxi to Tegel airport in Berlin, where Einstein and Feynman took a Lufthansa flight to Washington. Newton travelled to London with British Airways. Then Haller and Heisenberg drove to the main train station; Haller took the train to Bern, and Heisenberg left for Munich...

After the train had left the station in Magdeburg, it stopped suddenly. Haller woke up. He realized that he had slept and that he had been dreaming about meetings with Einstein, Feynman, Heisenberg and Newton. He went to the dining-car to drink a coffee.

Haller thought about his dream. He knew Heisenberg, since he had worked at the Max-Planck institute in Munich. With Feynman he had worked at Caltech, but he never met Einstein. He wondered how Newton would have thought about quantum mechanics.

About one hour later, the train reached the main station in Berlin. Haller took the subway to the Franzoesische Strasse, then he walked to the hotel "Gendarm". He had dinner at the steak house on the Gendarmenmarkt, sitting at the same table where he had been with Einstein, Feynman, Heisenberg and Newton a few days ago, in his dream.

Next morning, after breakfast, Haller went to the house of the Academy of Sciences in the Jaegerstrasse. At 9 a.m. the meeting of the Academy started.

Short Biographies
of the Physicists

Albert Einstein

Albert Einstein was born on March 14, 1879, the son of the merchant
Hermann Einstein and his wife Pauline. In 1880 the family moved to
Munich, where Hermann Einstein and his brother established a small

factory for electrical devices. Their company was the first to supply electrical light to the Munich Oktoberfest, as well as to a large part of the suburb Schwabing in Munich.

Einstein was a good pupil at school, particularly in the sciences. He read a lot, especially popular science books. He started school in 1885, one year after beginning to learn the violin, and from 1888 he went to the Luitpold high school in Munich.

At the beginning of the 1890s, the company of Einstein filed a petition for bankruptcy. The family left Germany and went to Milano. Albert stayed in Munich to finish his high school education, but he had problems with the school, and decided in 1894, to leave school without a diploma and to join his family in Milano.

Albert then applied for a place in the Swiss Federal Institute of Technology in Zurich. Since he did not have a high school diploma, he had to pass an examination which he failed. In the following year, he went to the high school in Aarau and obtained his diploma. He lived in Aarau in the house of the Winteler family. Mr. Winteler was Albert's high school teacher, and his son Paul later married Einstein's sister Maja. In 1896, Einstein began his studies at the Federal Institute of Technology.

He left the university in 1900 with a diploma as teacher of mathematics and physics. He applied for an assistant position at the Federal Institute of Technology in Zurich and at other universities in Switzerland, but without success. In 1901, Einstein became a Swiss citizen. In June 1902, he obtained a position in the patent office in Bern.

During his studies Einstein met his future wife, Mileva Maric from Serbia, who also studied at the Federal Institute of Technology. They married in 1902 and had two sons, Hans Albert (1904–1973) and Eduard (1910–1965). The family lived from October 1903 until May 1905 at Kramgasse 49, where today it is a small museum.

In 1905, at the age of 26, Einstein published some of his most important papers. In June 1905, he sent his paper "On the electrodynamics of moving bodies" to the Annalen der Physik. Afterwards he published his paper "Is the inertia of a body depending on the energy?" In this paper, the famous formula $E = mc^2$ appears. These two papers led to the theory of special relativity.

In 1909, Einstein received an extraordinary Professorship at the university in Zurich, and in 1911 he took up a Professorship at the German University in Prague. However, one year Later he returned to Zurich as professor at the Federal Institute of Technology.

At the beginning of 1914, Max Planck succeeded in convincing Einstein to come to the academy in Berlin. He became Director at the Kaiser-Wilhelm-Institute and professor at the University. Einstein published his theory of General Relativity in 1916. In 1919, Einstein was divorced from his wife Mileva, and shortly afterwards he married his cousin Elsa Löwenthal.

In May 1919 English astronomers, led by Arthur Eddington, observed in Brazil the light deviation of star light by the sun. The predictions of the general theory of relativity were confirmed. Suddenly Einstein was the most famous scientist in the world. He received the Nobel Prize in 1921, not for his theory of relativity, but for his work on the photoelectric effect in 1905.

In 1930, Einstein bought a piece of land in the village of Caputh near the city of Potsdam, where his summer house was built. During the following years Einstein lived in Caputh in the summer. In the autumn of 1932 he went to the United States, in particular to Caltech in Pasadena. He did not return to Germany since Hitler became chancellor in January 1933.

That same year Einstein became a member of the new "Institute for Advanced Study" in Princeton, where he lived at No. 112 Mercer Street. Einstein worked on a unified theory of gravity and electromagnetism, but did not succeed in finding the universal formula he was searching for.

In August 1939, shortly before the start of World War II, Einstein signed a letter, written by Leo Szilard, to the American President

Franklin D. Roosevelt, pointing out the danger of an atomic bomb built by the Germans. The Manhattan project was started, and in 1945 two atomic bombs were exploded in Japan.

In 1952, Einstein was offered the opportunity to become the president of Israel, but he did not accept. He died in Princeton at the age of 76 on April 18, 1955.

Richard Feynman

Richard Feynman was born on May 11, 1918 in Far Rockaway near New York and lived there until 1935, when he began to study at the MIT. In 1939, he went to Princeton in order to work on his Ph.D. in the group of John Wheeler. In his thesis Feynman developed the path integral formalism in quantum mechanics, following earlier ideas of Paul Dirac.

After finishing his Ph.D., Feynman, like many of his colleagues, went to Los Alamos to work on the Manhattan project. He witnessed the first explosion of an atom bomb near Alomogordo in New Mexico in Summer 1945. After the end of the war Feynman got a Professorship in theoretical physics at Cornell University in Ithaca, which he left in 1951 to take up a Professorship at the California Institute of Technology.

In 1942, Richard married his girlfriend Arlene Greenbaum, who died of tuberculosis in 1945. With his third wife Gweneth he had two children, Carl and Michelle.

Feynman worked in many areas of modern physics, especially in quantum electrodynamics, nuclear physics, particle physics and solid state physics. In 1965 he received the Nobel Prize, together with Julian Schwinger and Shinichiro Tomonaga, for his contributions to quantum electrodynamics. Feynman was one of the leading physicists of his time. His strength was his ability to reduce very complex problems to a few fundamental features. He found that many quantum phenomena in quantum electrodynamics could be described by diagrams, which are today called Feynman diagrams. The "Feynman Lectures" in physics are famous indeed.

In 1968, Feynman was able to explain the interesting results found at SLAC in Stanford, in his parton model of the hadrons. Later it was found that partons are nothing but quarks and gluons in the theory of quantum chromodynamics (QCD). Feynman was strongly interested in this theory, and until about three weeks before his death he gave lectures on QCD.

Richard Feynman died of cancer on February 15, 1988 in Los Angeles.

Werner Heisenberg

On December 5, 1901 Werner Heisenberg was born in Wuerzburg in Northern Bavaria. In 1910 his father, August Heisenberg, obtained a Professorship in philology at the University in Munich. Werner Heisenberg passed his high school examination in 1920, and afterwards studied physics at the same university. In 1923, he completed his Ph.D. thesis, advised by his professor Arnold Sommerfeld. He then went to the university in Goettingen and worked in the group of Max Born. One year later, he received an invitation from Niels Bohr to go to the institute in Copenhagen.

At that time nobody understood the theory of atomic phenomena. Heisenberg and his friend Wolfgang Pauli, who was also a student of Sommerfeld in Munich, were among the first who requested that the concepts of atomic theory, introduced by Niels Bohr and Arnold Sommerfeld, should be abandoned. Heisenberg started to work on a new formalism for quantum mechanics. In June 1925, Heisenberg traveled to the Island Helgoland in the North Sea, in order to avoid hay fever. He found that his new formalism could describe the atomic properties very well, and thus the new quantum mechanics was born.

In 1927, Heisenberg took the position of assistant to Niels Bohr. He discovered that in atomic phenomena an uncertainty was always present. It was impossible to measure the location and the momentum of a particle at the same time. The uncertainties obeyed an uncertainty relation: $\Delta p \times \Delta q \approx h$. Thus in the atomic theory there is no strict causality, like in classical mechanics. The new quantum theory was a theory about probabilities. In particular this was proposed by Max Born, Heisenberg's professor in Goettingen.

In the autumn of 1927, Heisenberg, only 26 years old, was offered a Professorship at the university of Leipzig. Whilst there, he worked in particular on the relativistic extension of quantum mechanics. Together with Wolfgang Pauli he developed in 1929 relativistic quantum field theory, in particular quantum electrodynamics. After the discovery of the neutron in 1932, he started to work on the theory of the atomic nuclei. He realized that a new short-range interaction, the strong interaction, must be responsible for the forces between the nucleons, the protons and neutrons.

In 1932, Werner Heisenberg received the Nobel Prize for the formulation of quantum mechanics at the age of only 31.

In 1942, Heisenberg accepted a position at the Kaiser-Wilhelm-Institute and at the university in Berlin. There he worked mostly on nuclear reactors. After the war Heisenberg was forced to stay for several months at Farm Hall in England. There he heard the news about the American atom bomb, dropped in Hiroshima, and expressed surprise that the American physicists were able to build such a bomb.

After the war Heisenberg was very active in building up science in Germany. He was the first director of the new Max-Planck-Institute for physics in Goettingen. At the end of the '50s, this institute was moved to Munich, and he worked mostly on his nonlinear spinor theory for elementary particles.

Heisenberg was also one of the founders of CERN near Geneva. As president of the Humboldt foundation, he invited young scientists from other countries to effect long-term research stays in Germany. Werner Heisenberg died on February 1, 1976 in Munich.

Isaac Newton

Isaac Newton was born on January 4, 1643 in Woolsthorpe (England). He went to school and afterwards to the high school in Grantham near Woolsthorpe. Afterwards Newton studied at Trinity College in Cambridge. The professor of physics in Cambridge, Isaac Barrow, recognized that Newton was outstanding in research and promoted him. Newton worked both in mathematics and physics, and discovered the differential calculus and the equations of classical mechanics. In 1669, at the age of only 26 years, Newton was offered the chair of Barrow.

Newton ranks amongst the most important scientists in history. He made fundamental contributions to classical mechanics, optics, mathematics and chemistry. However, his contributions to physics are more important than those he made to other fields.

In his book "*Philosophiae Naturalis Principia Mathematica*" Newton published the law of gravitation. He introduced the first natural constant in physics, Newton's constant G of gravity. He was able to show that the laws of Johannes Kepler followed from his law of gravitation. He introduced the concepts of absolute time and space. In 1905, Albert

Einstein discovered that these concepts had to be replaced by relativistic space-time.

Isaac Newton died on March 31, 1727 in London. His final resting place is in the Westminster Abbey.

INDEX